AMERICAN MANUFACTURING
...A STORY OF SURVIVAL AND SUCCESS

BY DR. JERRY L. SHOUP

AMERICAN MANUFACTURING
... A Story of Survival and Success

Copyright © 2018 Jerry Shoup
All rights reserved.

No part of this book may be reproduced in any form or by any electronic or mechanical means including information storage and retrieval systems, without permission in writing from the author. The only exception is by a reviewer, who may quote short excerpts in a review.

Author Contact: shouppurebook72@gmail.com

ISBN: 978-1-9466710-8-0 (Paperback)
ISBN: 978-1-9466710-7-3 (E-book)

Library of Congress Control Number: 2018937624

Printed in the United States of America

To Brad
Thanks for coming
Jerry Shoup
May 19, 2018

TABLE OF CONTENTS

Foreword *5*
Acknowledgments *7*

1. Senior Research Technologist 11
2. Manager West Coast Pickle Operations 27
3. Holland Shoe Factory 37
4. Manager of Technical Services Pickles, Vinegar, and Sauces 41
5. Consolidation of Midwest Grading and Brining into Holland 53
6. Holland Factory & Agriculture Manager 57
7. 100-Year Anniversary of Heinz in Holland 69
8. Adversity is a Great Motivator 75
9. Perrigo 81
10. Co-pack of Competitor Pickles 85
11. Project Millennia 89
12. Transfer of Vinegar Bottling and Mustard, Worchestershire, Horseradish and Jack Daniels Barbeque Sauces into Holland Factory 93
13. Swanson Pickle Company Contract 95
14. Vlasic Acquisition Project 99
15. Holland City and Heinz Cooperation and Mutual Respect 103
16. Co-pack Opportunities India and Mexico 107
17. Heinz Waterfront Walkway 113
18. Heart Power 121
19. History Channel Modern Marvel Program on Acids – Vinegar 127
20. Delmonte Spinoff 129
21. Private Label Pickle Sales Manager 131
22. Millennia Impact Upon Holland 133
23. Comments from Outside and Inside Heinz 135

Afterword *141*
References *143*

FOREWORD

I was extremely fortunate to have worked for the H. J. Heinz Company for 38 years, from 1974 until my retirement in April 2013. I am proud to have worked for a great company. The history of the company and its founder, Henry J. Heinz, has long inspired me.

Mr. Heinz advocated for pure foods from the beginning of the company, in 1869. He guaranteed that his foods were made pure and did not contain any toxic materials like those frequently used by some of his competitors to preserve their foods. He was a strong advocate for the Pure Food and Drug Act of 1906 which led to the FDA.

Mr. Heinz was also committed to the health and well being of all of his employees. Mr. Heinz first offered life insurance in 1917, over concerns for employees fighting in WWI. Heinz may have been the first company to do so.

One of his guiding principles was what he called "heart power." He defined this as doing everything possible to make employees feel good about the workplace. A book written about Mr. Heinz called him "The Good Provider." The Heinz history served as a model for me. I tried to emulate the Heinz values as best I could during my career. The first Heinz product was horseradish. Pickles, vinegar and ketchup as well as several others followed.

Pickles were the founder's favorite product and Genuine Dills his favorite pickle. Mr. Heinz made his Genuine Dills in 50 gallon barrels with his Pure Food Logo on the barrelhead. His Pure Food Logo is on the front cover of this book. Today Genuine Dills are produced in large 12,000 gallon tanks.

Heinz was the #1 pickle brand for many decades but brand share started to erode in the late 1960s and early 1970s. The competition from Vlasic and other competitors was intense. Heinz started to invest in the tomato ketchup business and in other products, which were more profitable than pickles. Spending the money necessary to defend the pickle market share did not make sense for Heinz.

Heinz souring on the pickle business became a cloud over the Holland, MI factory which at one point claimed to be the largest pickle factory in the world. Heinz Holland is the only company in the history of Holland to span three centuries under the same ownership and product line from 1897 to 2013.

Heinz headquarters tried to close the Holland factory several times during my career through co-packing and selling the facility. This book emphasizes the commitment and bond that all employees at the Holland factory had, and the steps taken to convince Heinz to not only keep the factory open but ultimately to expand it by transferring various Heinz products from other Heinz factories to Heinz Holland, MI.

ACKNOWLEDGMENTS

My wife, Kathy who was always there for our three boys and maintaining our home when I was not.

My dad, Roy, who was my coach and my mom, Dorothy, both of whom worked so I could graduate debt free.

My Heinz bosses, Dan Nolan and Dan Poland, whose leadership, encouragement and support were critical.

Tracy, CA factory pickle department support staff members Robert Byrne and Mike Fletcher, who helped me manage the department. Ogden Perry, Tracy Factory Manager, who supported me, Bruce Sharp in Tracy and Holland, who kept operations maintained and running, Isleton tank yard Office Manager Shirley Davidson and Supervisor Randy Arroyo.

My Holland factory staff: Larry Heckel, Ed Balint, Dale Weigel, Bill Snyder, Terry Prins, Mike Mooney, and Shannon Chada, whose dedication and support kept the Holland factory from failing.

Terry Prins, Manager of Engineering retired, for identifying needed corrections to the book text.

Department managers Kurt Schonfeld, Joe Kuzmanko, Ryan Hunderman, Evan Dawdy, Phil Wesorick, and all other management and RWDSU team members who contributed to the success of the Holland factory.

My factory manager colleagues Jerry Kozicki, Kish Patel, and Tom Green, who were very supportive of the Holland factory and me and who served as models for me.

Randy Vande Water, author of "Heinz Holland: A Century of History," which enlightened me concerning the history of Heinz in Holland.

Joanne McCauslin, for the wonderful album of pictures of my years at Heinz Holland, many of which are in this book.

Joshua Wise and other members of Luana Wise's family, who gave me permission to talk about her and her grave marker.

Robert Dykstra, Holland concerned citizen and Rick Jenkins, Holland Museum Registrar, who enlightened me concerning the history of the Holland Shoe Factory.

Don and John Swanson, from Swanson Pickle Company, who trusted me and the Holland factory to expand brining and grading operations, which helped keep the Holland factory viable. Katie Hensley, from Swanson Pickle Company, who provided pictures of "bloating" and purging.

Al McGeehan, former Holland mayor; Nancy De Boer former Holland Councilwoman and current Mayor; and Soren Wolff, former Holland city manager; who partnered with the Holland factory and me to support operations and the Heinz Waterfront Walkway.

Professors and researchers, including Dr. Henry Fleming and Dr. Roger McFeeters of North Carolina State University, Dr. Ralph Costilow of Michigan State University, and Dr. Ron Buescher of the University of Arkansas. Their research and presentations helped me understand "bloating" during fermentation, the role of calcium in firmness, side arm purging, and brine reuse.

Pickle Packers International and the Vinegar Institute, trade associations, which gave me information through scientific meetings and opportunities to work with key supplier and competitor contacts.

SENIOR RESEARCH TECHNOLOGIST 1974

I graduated from Wooster High School in Wooster, Ohio in 1964. I entered Bowling Green State University in Bowling Green, Ohio in pre-pharmacy in the fall of 1964.

During the next two years in pre-pharmacy, I decided that pharmacy was not the career I wanted to pursue. I worked during the summers at the Ohio Agricultural Research and Development Center in Wooster, which was part of The Ohio State University in Columbus.

I became interested in food processing and technology. In the fall of 1966, I decided to transfer from Bowling Green State University to The Ohio State University to major in Food Processing and Technology.

I graduated from The Ohio State University in August 1968. During the four years at Ohio State and Bowling Green State University, I had a 2S draft deferment as a full time college student.

Upon graduation my draft status changed to 1A, classified as ready for the draft. In December 1968, I received my draft notice. I decided to enlist in the U.S. Army. After basic and advanced training as a Communication Center Specialist, I was sent to Vietnam in June 1969. My first son was born in November 1969. I learned about his birth three days after his birth through the Red Cross. After returning from Vietnam, in June 1970, I decided to go to graduate school.

Jerry Shoup
Nha Trang Vietnam 1969

During this time the State of Ohio offered a Vietnam GI bill with 30 months of financial assistance. I was also given a graduate research assistantship in the food processing and technology department at Ohio State.

I wanted to complete a PhD project that would have a significant impact upon the food industry. Limiting chlorides in ground water was a major issue in the 1970s.

My project, "Salt Free Processing of Pickling Cucumbers," offered an alternative to brine fermentation of cucumbers. I stored cucumbers in 30 gallon drums with acetic, lactic, and citric acid solutions and a preservative to retard fermentation.

I withdrew samples periodically to determine the quality of the cucumbers, which were to be processed into hamburger dill slices. Only acetic acid, the main acid in vinegar, showed some promise as a storage acidulant.

I published a paper on this research in the Journal of Food Science, which resulted in the opportunity to present the paper at the Institute of Food Technologists Annual Meeting in New Orleans in May 1974. I did not know that Dr. Henry Fleming, one of the most well known scientists in the pickle industry, would be presenting a paper at the same meeting. The paper would cover bloater formation, which was the biggest problem facing not only Heinz but also the whole pickle industry.

I had no idea that my presentation at this meeting would change my life and set in motion a 38-year career, which I thoroughly enjoyed.

IFT Presentation Schedule

Dr. Richard Hein, Director of Research & Development for Heinz, was in the audience. After the presentation he walked up to me and asked if I would go to Heinz headquarters in Pittsburgh to further discuss this research. I accepted his invitation to go to Heinz headquarters.

I traveled to Pittsburgh prepared to discuss my research with a group of Heinz R&D employees, but suddenly it became apparent that I was being interviewed for a position. When I arrived at the Heinz headquarters, I found myself being shuttled from office to office to meet Heinz Research & Development personnel.

A few days after returning to my graduate school office in Columbus, I got a call from my graduate advisor, Dr. Wilbur Gould. He told me that Heinz wanted to offer me a position. I told my advisor that I was not looking for a job because it would take another year to finish the dissertation, complete my graduate committee exams, and defend the dissertation.

My advisor told me that I would be crazy not to take the position, that it was a great opportunity. He told me to negotiate with Heinz to get the time I needed to travel back and forth to Columbus. I took that advice. Heinz agreed to give me the time; they wanted me to complete the degree requirements.

I accepted the position, Senior Research Technologist, and my family and I moved to Waterville, OH, which was near Bowling Green near a Heinz Agriculture Research Facility. Dr. Ben George was my first boss and my mentor.

I did travel back to Columbus as needed to complete my degree requirements. I received my PhD in Food Processing & Technology one year later, in June 1975.

My first assignment was to manage a major research project, which the Heinz R&D group had developed, to determine the best cucumber brine fermentation procedure for maximizing quality and yields of hamburger dill slices.

The research project had a solid experimental statistical design with five replicates of various treatments. The goal was to determine the best brining procedures and the effect of cucumber grade size and the amount of handling of the cucumbers received prior to brine fermentation. The research was conducted in small 10-gallon tanks so that many different treatments and trials could be conducted efficiently.

The tanks needed to be located outside, exposed to sunlight, similar to large commercial size Heinz brine tanks. The ultraviolet light in sunlight controls the amount of film yeast that grows on the surface of brine tanks when they're shaded from sunlight.

Of major concern to the R&D group developing the research project was rain. It could affect the treatments in the small 10-gallon research tanks. Steve Lindquist, a good chemical engineer, was assigned to manage the project and procure materials temporarily until a research manager could be hired to take control of the project. Steve identified a special Plexiglas material, which could transmit ultraviolet light. Sufficient special Plexiglas material was ordered to cover all the 10-gallon tanks.

After I was hired and took control of the project, I challenged the use of the special ultraviolet transmitting Plexiglas research tank tops. I knew that only a narrow band of ultraviolet light in the middle of the ultraviolet spectrum was germicidal and could control film yeast growth. We took samples of the special Plexiglas to the physics department at Bowling Green State University for testing. We verified that the only UV light that passed through the special Plexiglas was at the upper end of the UV spectrum near the visible light spectrum and not in the germicidal band.

We decided to leave the research tanks uncovered and fully exposed to sunlight. We discussed that if we kept the research tanks full, topping off any daily losses of water due to evaporation, we could preserve the integrity of the research brines because any rain water would run off the top of the tanks. Any minimal effect of the rainwater would affect all tanks equally and therefore not be a significant problem.

Dr. Richard Hein, who attended my presentation at the Institute of Food Technologists, was the Heinz representative at the Pickle Packers International and the Vinegar Institute trade associations. Shortly after I took over the management of the Heinz brining research project, Dr. Hein retired from Heinz. I was named the new Heinz representative to these trade associations.

These trade associations were invaluable to me for the next 30 plus years. I was very active and attended most technical and scientific meetings concerning pickles and vinegar. I served as Board Chairman of the Vinegar Institute and President of the Pickle Packers International.

The technical knowledge and contacts with key suppliers and competitors in the pickle and vinegar businesses laid the foundation for many cost reduction opportunities during my career. These opportunities included improved brining procedures, co-packing opportunities for other pickle companies, and contracts between suppliers, which controlled factory costs. These opportunities, coupled with internal factory reorganizations and cost reduction projects, kept the Heinz Holland factory open despite significant pickle business volume declines.

In 1974 – 1975, during the spring and fall Pickle Packer meetings, I attended sessions with Dr. Henry Fleming and others from the USDA Food fermentation Lab at North Carolina State University and Dr. Ralph Costilow from Michigan State University. These sessions were critical for me in managing the Heinz brining research project. They helped me understand the mechanism of bloater formation in cucumbers during the brine fermentation process.

When cucumbers are immersed in a salt brine solution, the salt in the brine causes the natural cucumber sugars to diffuse out of the cucumber into the surrounding brine. Lactic acid microorganisms on the skin of the cucumbers start to grow, converting the sugars to lactic acid.

This fermentation process produces carbon dioxide gas. This gas is highly soluble, and in a dissolved state it diffuses inside the cucumber. As lactic acid is produced the pH decreases. Carbon dioxide is less soluble at lower pHs. At some point the carbon dioxide comes out of solution into a gas phase. This sudden release of gas can blow out the center of the pickle much like blowing up a balloon. The result is a pickle with a hollow center, which is known as a "bloater."

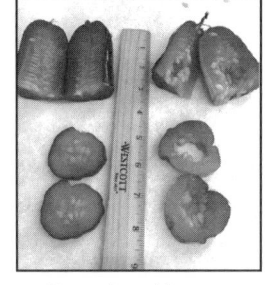

Purged no-bloating
Non-Purged bloated

During 1974-1975, Dr. Henry Fleming and others at the North Carolina State University USDA Food Fermentation Laboratory proposed purging the fermenting brine with nitrogen gas by bubbling the nitrogen through a coil tube at the bottom of the brine tank. They discovered that this could prevent bloating. Nitrogen gas is insoluble and therefore remains in a bubble state. The nitrogen bubbles rising to the surface agitate the fermenting brine solution much like shaking a bottle of a carbonated beverage causing the carbon dioxide to foam and come out of solution. The carbon dioxide can then be discharged at the top of the tank into the atmosphere. This continual purging prevents the dissolved gas from building up and entering the cucumber.

I was very excited about this research at North Carolina State. I included treatments purging the brine during fermentation in the 10-gallon research tanks in the Heinz brining research project. These purging treatments proved that purging was the solution to preventing bloating. Bloating was such a problem and huge financial loss for the Heinz Holland factory in 1974 that Ed Schierbeek,

Edward Schierbeek

Holland Factory Manager, wrote to his boss, Bill States, General Manager of Manufacturing at the Heinz Pittsburgh headquarters, recommending Heinz exit the hamburger dill slice business in #10 pickle cans for McDonalds and other foodservice customers.

One can understand Ed's frustration when the #10 pickle cans were costed and priced based upon a line speed of 1,500 cases per shift. The factory was achieving only 60 cases per shift. Approximately 70% of pickles coming out of Heinz pickle tanks in Michigan were bloated, resulting in slices with hollow centers.

At the time the Heinz standard was no more than 10% quality defects, which included hollow slices. To meet the standard the line had to be slowed down and extra production workers scheduled. These workers would squeeze each pickle to remove the bloaters, since a bloated pickle would collapse and squirt liquid when squeezed.

When we heard about the letter recommending exiting the business, we convinced Heinz to retain the business because there was research pending which could eliminate the bloating problem. After fully implementing the new purging process, production yields increased to over 90%. We doubled the total cases per shift with fewer production line workers.

Prior to 1970 bloating losses were far fewer. We were perplexed by the brining bloater problem. What happened in the early 1970s to exacerbate losses due to bloating?

In the 1970s Heinz started working with MSU agricultural engineers and other machine harvesting equipment suppliers to develop machine harvesters for cucumbers. Heinz was one of the pioneers in the development of machine harvesting of cucumbers. Machine harvesting was absolutely necessary for the long term survival of the pickle business because hand harvesting of cucumbers was becoming increasingly problematic due to the availability and difficulty in hiring migrant farm workers.

Dale Marshall and other agriculture engineers at MSU and other universities started to focus on the increased handling damage to cucumbers by machine harvesting, and by the successive drops of cucumbers during conveying and the filling of brine tanks.

Cucumbers respire after harvesting, giving off carbon dioxide. Handling damage could exacerbate the respiration and amount of carbon dioxide produced just prior to brining. Could this contribute to bloating? Could the handling damage to the cucumber structure and integrity from machine harvesting, conveying and tank filling make the cucumbers more vulnerable to bloating? These questions could not be answered during the brining research but do seem plausible.

We designed treatments in the 10-gallon brining research tanks, placing cucumbers gently into the tanks and brining without purging. The results showed that these pickles with minimal handling damage did not have any significant bloating.

The various cucumber varieties also showed differences in the brining quality. To make machine harvesting viable, cucumber varieties needed to be developed which would result in higher yields per acre for the grower to justify the capital cost of machine harvesters. Heinz agricultural researchers developed the Heinz cucumber variety 3534. This Heinz variety did set a high number of cucumbers on the plant vine, which resulted in good yields for the grower. Unfortunately our brining research showed the quality of the brined Heinz 3534 cucumbers was lower than other non-Heinz varieties.

After two years of brining research in 10-gallon research tanks, the results proved that purging minimized bloating regardless of whether the bloating was due to the fermentation process or some combination of the fermentation process and handling damage due to machine harvesting. I believe that the new brining purging process and the development of cucumber machine harvesting were the greatest advances in the history of the pickle business.

In 1976 it was time to scale up the new purging process to commercial size Heinz tanks. My family and I moved from Waterville, OH near the Bowling Green Research Facility to Holland, MI.

Dr. Ralph Costilow from MSU advocated a "side arm purging system" for pickle tanks. This system included a four-inch PVC pipe attached to the sidewall of the tank. Flexible ¼ inch tubing was inserted into the four-inch pipe and attached to a porous diffusion stone positioned at the bottom of the tank. Dr. Costilow recommended using air as the purging gas. Air was pumped under low pressure through the ¼ in. tube down through the diffusion stone, resulting in small bubbles, which would lift and agitate the

Side Arm Purge Tube

Tank Undergoing Purging

fermenting brine and discharge the brine into a two-inch horizontal pipe extending to the opposite side of the tank. This resulted in a continual pumping of the brine and discharging the carbon dioxide at the surface of the tank. This proved to be the most efficient way to purge the tank brine. Heinz and most other pickle companies adopted this "side arm purging system."

For the next two years I supervised the scale up of the side arm purging process from 10-gallon research tanks to over 700 12,000-gallon Heinz tanks in Michigan. This included laying out the tubing and PVC purging gas distribution systems and training the brining employees at the various tank yards. Long-term tank yard brining employees were very concerned since the new procedure would significantly change the only way they knew to brine cucumbers. I was told I better know what I am doing. I told them to trust me.

I had always considered myself to be a technical R&D person doing research. The interaction with maintenance and production personnel at the various tank yards during the purging installations and training activities was personally enjoyable. I started to think that working with people and managing operations is the direction I should go in my career rather than the technical research pathway.

My doctorate research concerning salt free acidulant storage of cucumbers was one of the main reasons Heinz hired me. Ironically I never pursued this research during my Heinz career. Twenty five to thirty years later, companies in India, Mexico, and Central America started to use a slight modification of this process to sell cucumbers known as acidified gherkins to customers in the U.S. including Heinz Holland.

Heinz employees hired in 1974 and 1975, including myself, were known as Heinz pacesetters. The pacesetter group members were supposed to be the future leaders of Heinz. I attended pacesetter seminars discussing Heinz values and goals. Looking back, some years later, I realized that there were only a few Heinz pacesetters remaining.

I believe Dave Willard was one of the pacesetters hired at the same time I was hired. Dave was assigned to Heinz Holland to observe operations and work with Heinz Holland to make productivity and operational improvements. During this time Ed Schierbeek, Heinz Holland Factory Manager, announced his retirement after serving 40 plus years with Heinz. Dave was named the new factory manager.

Ed was well known and highly respected within Heinz. A special retirement dinner and program was organized to honor Ed for his years of service. The President of Heinz attended. Unfortunately, when the President arrived he noticed that Hunt's ketchup was on the table. The President was so upset that he called the Heinz regional sales manager and made a big issue concerning Hunt's ketchup being on the table in a city with a large Heinz factory.

I learned a key lesson that night. For the next 30 plus years whenever I was involved in organizing a retirement or business dinner with Pittsburgh headquarters executives attending, I made sure Heinz Ketchup was on the table. If the restaurant did not normally have ketchup on the table for business dinners or events, I brought the ketchup and insisted that the ketchup was on the table.

The Vlasic pickle company was growing significantly, taking pickle market share from Heinz. Vlasic hired Dave Willard. Dave was aware of my salt free acidulant storage research. He invited me to go to Vlasic headquarters in West Bloomfield, MI and tour the headquarters. Vlasic discussed their growth plans and wanted me to join the company. They offered me a significant position in the headquarters group. Perhaps Vlasic wanted to make sure Heinz did not use the research.

My wife and I discussed the offer. We liked living in Holland and did not want to move to West Bloomfield. Also I was very proud to be working for Heinz given the values and history of the company. We turned down the offer.

Sometime later, Dave left Vlasic and went to the Perrigo pharmaceuticals company in Allegan, MI. I did not know at this time that Perrigo would offer a key opportunity for Heinz Holland years later.

Upper management was perplexed by the declines in the pickle business. A decision was made to hire a business management consulting firm to study the Heinz pickle business. If I remember correctly the firm was McKinsey and Company business consultants.

As I remember, McKinsey concluded that the Heinz pickle business needed to be managed separately from the other Heinz brands due to the complexity of the business. Most other pickle companies were regional and only in the pickle business.

As a result of the study, a Pickle division was established. At the time, Heinz owned a private label pickle business in Henderson, North Carolina called Perfect Packed. Management of the Heinz pickle division was moved out of the Pittsburgh headquarters to Perfect Packed. Heinz hired Don Heineman from McKinsey to manage the new division. This ruffled many feathers in Pittsburgh.

The new purging process in Michigan was such a huge financial and operational success that Don Heineman wanted an appropriation to install purging in the Perfect Packed tank yard. He called me and asked me to come down to Henderson, NC to discuss the project with him. During the meeting he told me that Charlie Roberts, VP of Finance, in Pittsburgh needed to approve the appropriation. Don said that Charlie hated him and would never approve the appropriation if he contacted him. He asked me to fly to Pittsburgh and meet with Charlie to gain approval. I did meet with Charlie Roberts and he signed the appropriation for Perfect Packed.

A year or two later Heinz decided to consolidate private label pickles into Holland, MI. Perfect Packed was closed. Key management and production personnel were transferred from Henderson to Holland.

The next major business change which affected the pickle business and me was the re-organization of Heinz into the Consumer Products and Packaged Goods Divisions. These divisions would have separate management groups and different business focuses. The Consumer Products Division included the growth brands such as Ketchup and 57 Sauce and would be managed for growth. The Packaged Goods Division included stable brands which were not growing such as Pickles, Vinegar, and Soup; these needed to be managed to control costs and maintain profitability.

At this time there were five major factories in Heinz USA as well as other smaller factories. The major factories included Pittsburgh PA, Fremont OH, Holland MI, Muscatine IA, and Tracy CA. Consumer Products managed the Fremont and Tracy factories. Packaged Goods managed Pittsburgh, Holland, and Muscatine.

Packaged Goods was concerned about the Tracy factory, which was the major tomato processing operation for Heinz USA. The Tracy factory made ketchup and shipped tomato paste east for ketchup bottling. Tracy also produced some pickles. Southern California was a significant Heinz pickle market.

Tracy was a Consumer Products factory. Pickles were a Packaged Goods product. Packaged Goods realized that pickle production and cucumber agriculture procurement were a small part of the overall west coast operations and responsibilities. In Tracy, pickles would not get the focus necessary to optimize the business. This realization was consistent with the McKinsey consulting study, which recommended separating the management of the pickle business from the other major Heinz businesses.

Packaged Goods decided they needed their own manager of pickle production and agriculture procurement in the Tracy factory and Isleton pickle tank yard. Isleton was also the remaining Heinz tank yard that needed to be converted to the new purging process.

I was asked to move to California to manage the pickle production department inside the Tracy factory, the Isleton tank yard, and to lead the conversion of Isleton to the new purging process. I was also asked to procure cucumbers for both Tracy and Isleton.

I realized that accepting this offer would be a new adventure for me because I did not have any experience managing a pickle production department or procuring cucumbers. I accepted the new position and so my family and I moved to Stockton, California.

MANAGER WEST COAST PICKLE OPERATIONS 1978

The Manager of West Coast Pickle operations was an unusual position for Heinz and me. I was responsible for the pickle production department in the Tracy factory without having the support functions reporting to me. These support functions included financial controller, quality manager, maintenance manager, and human resources manager.

All the Tracy support functions reported to Ogden Perry, Consumer Products Tracy Factory Manager. I reported to George Daily, the Packaged Goods Factory Manager in Holland, MI. The Tracy factory manager reported to Burt Kleinsmith, the Consumer Products West Coast General Manager of manufacturing. Burt was responsible for both Consumer Products factories in Tracy and Stockton, CA.

I was also responsible for procuring all west coast cucumbers for pickle production in the Tracy factory and brining in the Heinz Isleton, CA tank yard. I was also responsible for Heinz cucumbers brined at SMS briners near Stockton, CA.

Prior to my arrival, all produce, including tomatoes, cucumbers and other vegetables, was procured by Chuck Bailey. He was well known and highly respected by tomato growers and food processors on the west coast. Chuck had a huge responsibility procuring most of the tomatoes for Heinz Ketchup bottling for all factories.

Consumer Products manufacturing and agriculture management personnel were bewildered by the fact that Packaged Goods sent a scientist to manage pickle operations on the west coast. I heard many comments about this.

Soon after my arrival at Tracy factory, Ogden Perry called me into his office and gave me a stern warning. He told me that he would not be responsible for any unsatisfactory California state food inspections of the pickle department. This was not the welcome to Tracy factory I was expecting, although I soon learned why he gave me this warning.

The California state food processor inspector, Christine, was notorious among food processors. She had a reputation for threatening to shut down operations during inspections. When she became upset during inspections, one of her eyes would twitch. The Tracy pickle department had scored lower than other departments during Christine's inspections.

I can understand why the pickle department scored lower than other departments. It was apparent that maintenance spending was a lower priority in the pickle department. The department needed to be painted with some more lighting to make it brighter. Pickle department production workers had fabricated a make shift chute with conveyor belts to remove pickles from desalting tanks and convey them to filling lines.

Bob Byrne and Mike Fletcher were the department supervisors. They were excellent supervisors. I was amazed by the production numbers they achieved despite minimal maintenance spending. Dick Staas, the production manager at the Tracy factory, offered Bob a promotion in another department. I was thankful that Bob turned down the opportunity.

In my new position I had the authority to spend necessary funds to get the department painted with new lighting. We worked with outside contractors to fabricate stainless steel chutes which could be lengthened or shortened as needed to efficiently convey desalted pickles to production lines. The productivity improved and pickle department employees appreciated the improvements we made. It proves that a little heart power can go a long way. We were very proud of the fact that Christine gave the pickle department a high score during the next inspection.

Bruce Sharp was a maintenance supervisor at the Tracy factory who came up to me and congratulated us for the improvements in the pickle department. Bruce was an excellent maintenance supervisor who went the extra mile in addressing our maintenance needs. As a result, sometime later, Bruce was offered a promotion and moved to the Holland factory.

Ogden Perry was very supportive of me. We developed a good working relationship and he made sure that the various support functions provided the support we needed. I had great respect for Ogden.

Chuck Bailey explained to me that he procured all the cucumbers from M J Rosenmayer and SMS briners because the quality and customer service from SMS were excellent. They were great, dependable suppliers, which allowed him to focus all of his attention on tomato procurement. Chuck recommended that I continue to use SMS briners and M J Rosenmayer for cucumber procurement. Lou Rosenmayer of M J Rosenmayer is a food broker who supplies cucumbers to many pickle companies. He is well known and highly respected in the pickle industry. But Lou, representing SMS briners, was not the only potential supplier of cucumbers in the Tracy area.

Soon after I arrived in Tracy I got phone calls from Bill Straub from Straub Pickle Company in Stockton and Don Zeller from California Briners near Stockton. These potential suppliers found out there was a new Heinz pickle man at Tracy. Both thought they may have a chance to sell cucumbers to Heinz. I agreed to meet with Bill Straub and Don Zeller. I thought it made sense to increase competition for sales to Heinz and hopefully get some lower cucumber prices. Certainly I had the time and ability to focus upon the cucumber procurement, unlike Chuck who was totally focused upon tomato procurement.

I was concerned about the fact that Heinz did not have any experience procuring cucumbers from Straub or Zeller. I had several meetings with them and inspected the Straub pickle factory and their cucumber grading and brine tank yard. I inspected the California Briners cucumber grading and receiving facility. Don Zeller was primarily a cucumber grower who supplied cucumbers to Straub and other companies.

I met with Lou Rosenmayer and notified him I was considering splitting the cucumber procurement between the three suppliers. Lou was upset about potentially losing volume but did not give me the feeling that this would be a very risky move.

The final cucumber pricing from Straub and Zeller was lower than that of SMS, therefore the benefit of splitting the cucumber procurement between the three suppliers outweighed the risk and made sense to me.

In a meeting with Bill Straub, he agreed to brine our cucumbers in his tank yard, saving freight to Isleton. Bill Straub was aware of the new brine purging process and he knew I had been involved in installation of this process at Heinz tank yards. He said I know that you would want to use this process here if you want to brine your pickles here and we would like to learn the process.

The Straub brining cost per ton was favorable to the Isleton cost. Eliminating the cost of shipping cucumbers from Straub to Isleton and then shipping brined pickles from Isleton to Tracy for pickle processing would be a savings opportunity. Also, Straub, in Stockton, was closer to Tracy than the Tracy factory was to Isleton. This would further increase the savings opportunity of brining and shipping pickles from Straub to Tracy as compared to Isleton to Tracy. In the days that followed, an unexpected opportunity developed which solidified the decision to split the cucumber procurement.

I got a call from the General Manager of Lindsey Olive Company asking if we could meet for lunch to discuss a proposal. I was very intrigued by this call and wondered what proposal an olive company could make to Heinz.

During lunch he explained that every other year is a bumper crop for olives. That year the bumper crop had been huge and Lindsey did not have enough tank yard brining space to brine the crop. Would it be possible to use our brine tank yard in Isleton to brine olives? This was a one-year request, since the following year Lindsey would have time to insure the company had sufficient ongoing space.

After thinking about it for a few minutes I responded, "What you are asking me is if we would rent our pickle brine tank yard to you for your upcoming olive crop and then find brining space elsewhere for our Heinz pickles." He said, "Yes and we hope you will consider this proposal."

Knowing that Bill Straub had already offered to brine our cucumbers, I started to think that this could be a real income opportunity for Heinz and the Heinz west coast pickle operations. I told Lindsey that they would need to give Heinz $100,000 up front for rent and pay all expenses associated with brining the olives, including removal of the spent olive brine with the olives by the start of the next brining year. Also they would need to provide supervision of the brining at Isleton since we did not have any experience brining olives. They would need to accept the quality of the olives, holding Heinz harmless for any losses as long as Heinz brined according to Lindsey instructions. I remember leaving the lunch thinking this will never happen. To my great surprise, a few days later, I was given a check for $100,000 along with a rental contract including all the requirements I had specified. I had the contract reviewed by the Heinz Corporate law office and given a green light to move forward.

Joe Dague was the financial controller of the Tracy factory. I will never forget the concerned look on Joe's face when I brought him a check made out to H. J. Heinz for $100,000. I was the new guy in town and Joe did not know me very well. He probably thought there must be something shady here. He wanted to know the reason for the check. I explained it to him and then said I wanted the check to be credited to Heinz Isleton's overhead budget. He told me he would talk to Pittsburgh finance and get back to me. Heinz did cash the check. Isleton's total overhead budget for that Fiscal Year was approximately $300,000. Isleton had a very favorable cost year.

I needed to spend time with Shirley Davidson, manager of the Isleton tank yard, and Randy Arroyo, tank yard supervisor, to explain why we were not brining Heinz cucumbers at Isleton for the upcoming year. They were dedicated Heinz employees and concerned about what this would mean for the future. They finally realized this was a one-year change and an opportunity to make significant income for Heinz.

Packaged Goods scheduled a business plan review in Pittsburgh. I was so excited to go back to headquarters and discuss the pickle plans for the west coast.

This was my first trip back to Pittsburgh from San Francisco airport. The presentation went well and I felt really good flying back to San Francisco.

At Tracy, we filled five-gallon pails of hamburger dill slices for Golden State, a foodservice broker that purchased five-gallon pails for McDonald's on the west coast. I got a call from the McDonald's Quality Manager who wanted to visit the Tracy pickle department.

When the Quality Manager arrived I learned that she just wanted to visit because the quality of the Heinz hamburger slices improved so dramatically that she needed to understand what drove the improvement. I explained the new purging process. She was impressed and appreciated the information. Unfortunately we were undercut in price some months later and lost the McDonald's contract.

The West Coast Pickle Operations did well over the next few years until 1982. Owens/Brockway, Heinz glass supplier, announced they no longer had capacity to supply Tracy pickle glass from their Oakland, CA glass plant. West coast demand for their beer glass had increased dramatically. They announced that pickle glass for Tracy would need to be shipped from their Midwest plant, which served the Holland factory.

This was a death sentence for the Tracy pickle department. It did not make financial sense to ship empty glass from the Midwest for pickle production on the west coast. It made more sense to produce in Holland and ship the finished goods directly to Tracy. The Tracy pickle department was closed and west coast pickle production was moved into the Holland factory. My family and I moved back to Holland, MI. I accepted a new Technical Services position.

It was difficult for my family living in Stockton California for four years from 1978 to 1982. We had three boys ranging in age from three to nine. In Michigan we had good-sized lots with big yards. The boys had room to play and we lived in neighborhoods with other boys their age. In California the lots were small with fences around the backyards. There was not enough room to play. We would get phone calls from neighbors complaining if one of our boys happened to accidently throw a ball into the neighbor's pool. Our boys could not find any other kids their age to play with on our street in Stockton. In Holland, their friends lived next door. Fortunately, Bob Byrne, who worked with me in the Tracy pickle department, lived only a block away from us in Stockton. Bob and his wife Pat were originally from Holland and worked for Heinz

Holland. They moved from Holland to Tracy a year or two before we moved to Tracy.

Bob and Pat also had three children. Their son Robby was the same age as our son Stephen. Their daughter Chrissy was the same age as our son Ryan and their youngest son Shawn was the same age as our son Greg.

Pat and Kathy and were best friends, babysitting for each other whenever needed. The six children were best friends and played together daily. Pat and Kathy would take them to Mickey Groove Park and Zoo, which was near Stockton. Sometime later, Bob was offered a new position at Heinz Holland as supervisor of the vinegar generation department. He accepted, and the Byrnes moved back to Holland.

My parents visited us for a week and decided to move to Stockton to be near us. It was good for Kathy to have my parents in Stockton especially after the Byrnes left Stockton.

Kathy was the real hero in my story. I was too wrapped up in my Heinz positions in California and Michigan to spend much time with our sons. She was the one who raised our sons and she did a good job, given that they grew up to be fine young men.

Stephen is now a mechanical engineer in the automotive industry. Ryan received an appointment from Congressman Fred Upton to the Air Force Academy. He graduated in the upper 10% of the electrical engineering cadets. This enabled him to earn masters and PhD degrees in electrical engineering while in the Air Force. He now works for MIT's Lincoln Laboratory doing defense research. Greg works for Kraft/Heinz in Holland, Michigan.

3

HOLLAND SHOE FACTORY 1901

When I arrived back in Holland, I moved into a second floor office in the Heinz Agriculture administrative building at the corner of 15th street and Cleveland Avenue, which was adjacent to and in front of the original Historic Holland Shoe Factory.

1901 Holland Shoe Factory

The office had a panoramic view of the factory and tank yard where many of my experiments were conducted. The Holland Shoe Company was founded in 1902 to utilize an abundant supply of sole leather generated by the

Cappon & Bertch Tannery. Within a year the Holland factory was one of the busiest factories in Holland, employing 200 people and producing 1,000 pairs of shoes daily. By 1914, the shoe factory had doubled to 400 employees producing 3,000 pairs of shoes daily. In 1939, the Holland Shoe Factory merged with the Racine Shoe Company in Wisconsin, forming the Holland-Racine Shoe Company.

During World War II the Shoe Company joined the war effort, producing 240,000 pairs of shoes for the Navy. In 1964, the Holland-Racine Shoe Company was sold to a St. Louis firm. In 1967, the factory closed, as did many other shoe manufacturers across the country; they could not compete with low cost imports.

After the closure of the shoe factory, Heinz acquired the property, including the shoe factory building, and used the building and property to receive and grade farmer loads of cucumbers.

In the 1980s, Heinz consolidated the receiving and grading of farmer loads of cucumbers and out-sourced these operations to Swanson Pickle Company in Ravenna, MI. The shoe factory building was vacated and the surrounding property was used for storage. Some years later the shoe factory building slowly deteriorated and needed to be demolished. During the demolition, Robert Dykstra, a

1901 Shoe Factory Cornerstonea

concerned citizen whose mother and father worked for the Holland Shoe Company, alerted me that the original building cornerstone was in danger of being reduced to rubble with the rest of the building. We retrieved the original cornerstone and moved it to a safe area.

In discussions with former Holland Mayor Al McGeehan, it was decided to move the original cornerstone from the Heinz property onto city property adjacent to Heinz. This was necessary because Heinz was in the process of utilizing wooden pickle tank boards to build a perimeter fence along the Heinz Waterfront Walkway and remaining Heinz property. The original 1901 shoe factory building cornerstone is now located at the southeastern corner of Kollen Park drive and 16th street.

Holland citizens have asked what the cornerstone represents and have assumed that it marks the start of Kollen Park. The Holland Shoe Factory, which became the Holland-Racine Shoe Company was a significant part of the history of Holland. The history needs to be shared with citizens and visitors alike. I am now working with Andy Kenyon and Grace Smith from the City of Holland, Rick Jenkins from the Holland Museum, and Randy Vande Water, Holland historian, to design a prominent historical marker and a plaque with the history of the Holland Shoe Factory which can be seen from 16th street.

4

MANAGER OF TECHNICAL SERVICES PICKLES, VINEGAR, AND SAUCES 1982

After moving back to Holland from Tracy, I accepted a new position as Manager of Technical Services Pickles, Vinegar, and Sauces. As for sauces, I was responsible for only Worcestershire Sauce, better known in the halls of Heinz as "woozy" sauce. I reported directly to Jim Flynn, VP of Technical Services and Quality Assurance, for the Packaged Goods Division. Jim was located in our Pittsburgh headquarters.

George Daily was the Holland Factory Manager. George gave me all the support I needed. I had great respect for George.

I was responsible for new product development, current product improvement, and process development for all of these products. This was a very stimulating position for me, which I enjoyed.

Three product and package developers reported to me, all from their location in Pittsburgh. Jim Flynn and I discussed that, by being located in Holland where most of these products were produced, I could directly oversee new product factory batch runs for products and new package trials.

Chuck Johnston was the pickle product and process developer. Harold Queer was the packaging developer for glass bottles, cans, plastic bottles and flexible pouches. Mike McMahon was a chemical engineer and responsible for vinegar generation, processing and new technology.

Mike and I worked on converting the old white distilled vinegar beech wood shavings generators, which utilized a trickle down process, to the new submerged fermentation acetator process for vinegar generation. Our main focus was on optimizing the vinegar output at the various factories, primarily Holland.

A year or two later Mike moved into packaging development. This was a critical time for Heinz in the development of plastic squeeze bottles for ketchup and relish as well as for other Heinz varieties.

Chuck Johnston developed a premium line of pickles called "Heinz Old Fashioned Pickles." One of the pickle products, "Old Fashioned Bread n Butter Chunks," which included cucumbers cut into chunks with fresh onions and red peppers, was, I believe, the best pickle product Heinz ever developed. We also developed a new line of vinaigrettes in 12-oz decanter bottles. The honey mustard variety was my favorite. Unfortunately, Heinz did not put enough market spending behind these excellent new products and as a result they did not survive.

One new vinegar product we developed was Heinz Cleaning Vinegar, which was the only commercial product newspaper columnist Heloise ever endorsed at the time. I was able to welcome Heloise to Heinz Holland during the time Heinz Cleaning vinegar was being introduced. Heinz Cleaning Vinegar is still being sold today.

Heinz Genuine Dills was H. J. Heinz's favorite pickle product. This was a unique product which originally was flavored with whole dry spices in 50 gallon wood barrels, which needed to be rolled weekly to distribute spices and ensure the proper flavor.

Heinz bought the rights to launch Mr Yoshida grilling and marinading sauces for enhancing the flavors of chicken, pork and other meats during grilling and cooking. At the time Mr. Yoshida gourmet sauces were being sold in California club stores. I had the privilege of visiting with Mr. Yoshida while he demonstrated cooking with his gourmet sauces. We worked with Heinz headquarters product development and marketing groups to develop the packaging for the launch of these marinading sauces.

These were excellent sauces and we were very excited about this launch and the opportunity to produce these sauces at Heinz Holland. Early in the launch Heinz decided to pull the market spending for the launch and as a result this great opportunity for Heinz Holland ended. As I understand the reason for stopping the launch was it was decided that the timing was too early and not right for the introduction to the consumer of the use of these sauces for marinating.

In the mid-1900s making Genuine Dills in 50 gallon barrels was no longer cost feasible. Before I joined Heinz,

Genuine Dills were being converted to large wooden tanks which also required the conversion to spice oils.

I re-wrote the process sheet for purging the Genuine Dills with a side arm purger to pump and improve mixing of spice oils, all to improve the product and eliminate any bloaters. I felt it was critical to purge with pure nitrogen to ensure no issue with the spice oil flavors.

I was told early in my career by old timers in Heinz that you will never be successful converting to large commercial tanks. Heinz Genuine Dills have now been made in large commercial tanks for many years.

Two special assignment projects were personally and professionally stimulating for me. Dr. Fons Voragen from the University of Wageningen in the Netherlands contacted Heinz to determine if Heinz would agree to host him for a sabbatical for six months. Heinz contacted me to see if I could spend time with him. I looked forward to working with him.

Fons was a biochemist/enzymologist and an expert in apple processing. He had developed what he called an enzyme cocktail containing pectinases and cellulases which would result in enzyme liquefaction breaking down cellulose and pectins.

Cider vinegar is made by pressing apples extracting apple juice. After pressing and extracting the juice, the remaining solids or pomace must be discarded. Generally the pomace is fed to farm animals. The sugars in the juice are fermented by yeast to apple alcohol and the apple alcohol is converted to acetic acid in a vinegar fermenter resulting in cider vinegar.

We discussed the possibility that the enzyme cocktail could increase cider vinegar yields by liquefying all of the apple material including the pomace resulting in more sugars available for conversion to alcohol and then acetic acid vinegar. It was understood that L-glucose from the breakdown of cellulose would not be fermentable but liquefaction may result in more fermentable sugars extracted from the apple material as compared to mechanical pressing. It was an interesting hypothesis but did not prove to be beneficial.

Fons and his wife stayed with Kathy and I at our home. We drove around Holland with the Voragens to show them how the Dutch live in Holland, MI. We were amazed when Fons would see a Dutch name on a storefront and be able to tell us the province in the Netherlands where the family originated.

A biochemist from the medical school at the University of California at Davis contacted Heinz and announced that his research had identified the reason why vinegar was so healthful. He invited Heinz to come to his lab to discuss his research. Heinz asked me to go to UC Davis to discuss his research.

Acetobacter microorganisms are involved in the oxidation of ethanol to acetic acid vinegar. He had found significant amounts of Pyrroloquinoline Quinone PQQ in vinegar. According to the chemist, the anti-oxidant properties of PQQ are many times stronger than ascorbic acid/vitamin C.

We discussed that PQQ may be in the acetobacter cells and then released into the vinegar after the cells die and lyse. According to the biochemist, PQQ could be the next new vitamin if studies confirm the healthful benefits. He proposed rat studies feeding vinegar to rats and determining health improvements.

He wanted a considerable amount of money to conduct studies. We agreed to provide cider and white distilled vinegar for his studies.

The studies did show significant amounts of PQQ in both cider and white distilled vinegars. Heinz decided that it does not make sense to spend considerable funds for research to tell the consumer that vinegar is a healthy food when most consumers already know vinegar is healthy.

Grocery Spears in 24-oz. jars was a key pickle variety for Heinz Holland. We packed grocery spears with five Solbern spear packing machines. Each machine produced 800 cases per shift, resulting in 4,000 cases per shift. This productivity offered a cost advantage against some of our competitors who packed spears manually on large hand pack tables with 20 to 40 hand pack crews.

Initially the Solbern machines were designed to cut a cucumber into five spears and place the spears around the perimeter of the jar.

The circumference of the 24-oz jar dictated the cucumber diameter size needed for the machine to fill all the space around the perimeter of the jar. For the Heinz 24-oz jar, cucumbers needed to be 2 1/8 to 2 1/4 inches in diameter. This diameter range is within the industry cucumber grade size known as 3Bs.

This 3B diameter size can be a problem given the size of the seed cavity inside the cucumber. As a cucumber grows in size on the vine, the internal seed cavity matures with the cavity becoming softer and seeds larger. Many of the spears made from 3B size cucumbers appeared ragged with large seeds.

Vlasic and other competitors who were hand-packing spears into jars were able to utilize smaller cucumbers, such as grade size 3A which are 1 ¾ to 2 inches in diameter, with firmer seed cavities and smaller seeds.

Our Heinz pickle marketing group was upset and rightly so. Heinz was losing sales on the shelf due to ragged, poor spear appearance as compared to Vlasic and other competitors. This became such a problem that our quality assurance group included in the quality standard for 24-oz. spears a special grade C for poor spear appearance. This resulted in many holds of spear production for salvage sales at a 50% reduced price.

C Grade Spear Appearance

My boss, Jim Flynn, and I committed to fixing the problem. We initiated a project to develop a Solbern machine which would pack 3A size cucumbers to reduce cucumber size and improve appearance of grocery spears.

Good spear appearance

We met with Solbern to discuss the project. Solbern proposed modifying our current five spear packing machines from the current "5 cut machine" cutting two cucumbers into 10 spears and place packing the 10 spears around the circumference of the jar to a "6 cut machine" cutting two cucumbers into 12 spears and placing the 12 spears around the perimeter of the jar. Placing two additional spears with "6 cut" around the perimeter of the jar would take up more circumference which would allow reducing the diameter size range from the 3B (2 ⅛ in to 2 ¼ in) to an upper 3A to lower 3B (1 ⅞ in to 2 ⅛ in). This did improve grocery spear appearance on the shelf.

With the modification to six cut we were able to improve the shelf appearance of spears and still retain our productivity advantage with machine packing as compared to hand packing. The efficient Solbern machine packing line was critical for the Holland factory to not only pack spears for Heinz but also to co-pack for competitors. This generated income to offset future Heinz pickle volume declines, helping to keep the factory viable and the doors open.

The success of the grocery spear line at the Holland factory was due in large part to Terry Prins, who would be the future Manager of Engineering, reporting to me.

The Solbern spear packers were complicated machines to run and to maintain. From the beginning of the Solbern spear packer installation for the Holland factory, Terry took charge of the spear packing line, learning how to efficiently run and maintain the machines.

Solbern told me that Terry Prins was the most successful person in the pickle industry at optimizing spear packing machine production. Vlasic told me they did not install Solbern machines permanently because they could not find anyone who could make the machines perform properly.

For most of my Heinz career, I was involved with both the Pickle Packers Int. (PPI) and Vinegar Institute (VI) Trade Associations. I believe this involvement was invaluable to me in my career. I was able to keep abreast of the latest scientific and technical knowledge in the manufacturing of pickles and vinegar.

I was able to meet and interact with most company presidents and technical personnel. These contacts were the key reasons why Heinz Holland was able to co-pack for Steinfeld Pickle Company, Nalley Fine Foods and Paramount Foods. These co-pack opportunities helped sustain the Holland factory during declines in Heinz pickle market share and Heinz pickle production. Co-pack income was critical for a few years to offset the loss in fixed overhead absorption until we had time to implement other cost reduction ideas and diversify our production product portfolio.

The strategy at Heinz Holland was to hang on by controlling factory costs until other new products could come into Holland and stabilize the production volume. This is in fact what happened. Several new products eventually came into Holland, diversifying and increasing the profitability of the factory production portfolio.

I was able to meet and interact with all the key suppliers and academic researchers. I always knew the right person to contact if I had a question or needed advice concerning any aspect of the industries. This resulted in reciprocal factory visits and an understanding of how Heinz compared to many of our competitors.

I served on the Scientific Committees and Boards of the Associations for over 30 years. I served as President of Pickle Packers and Board Chairman of the Vinegar Institute.

When I was president of Pickle Packers, a huge opportunity developed when university scientists developed parthenocarpic or seedless cucumbers. The research concerning seedless cucumbers needed to be conducted in isolation from random pollination by bees. The research was conducted in huge green houses in Costa Rica. Pickle

Packers funded some of the research. The board traveled to Costa Rica to review the research.

I remember getting a call from Heinz antitrust attorneys. The attorneys informed me that Heinz had been named as a co-defendant in a Pickle Packers antitrust lawsuit. A small foodservice pickle packer filed a suit that members of PPI colluded to fix prices and hurt his business. The Heinz attorneys wanted to know my involvement in PPI.

I explained that my involvement was totally in the technical and quality improvement of the industry. I never met with any sales or marketing personnel or discussed pricing.

Every meeting that was held at PPI's annual spring and fall meetings included an antitrust lawyer from a Washington law firm. That lawyer gave an antitrust warning before the start of every meeting and stayed in the room for the entire meeting. A few weeks later, I was happy to hear that Heinz was found to not have any part in the suit and was dropped as a co-defendant.

Heinz was also named as a co-defendant in a Vinegar Institute lawsuit. In this case, a small food processor claimed that an employee was working around a mustard tank. He opened a manhole lid in the top of the tank and was overcome by vinegar fumes. He fell and died. The suit claimed that the company did not receive the proper hazard information accompanying the vinegar load to warn of the hazard. This suit was eventually dropped.

While I was factory manager at Heinz we received a citation from a MIOSHA inspector claiming Heinz Holland was in violation of the Process Safety Management, PSM, standard. The standard gives an exemption for ethanol stored under atmospheric conditions. The inspector claimed that pumping ethanol out of the storage tank into a receiver tank and, while pumping, immediately diluting the ethanol with water to a concentration which is non-flammable, is a processing step. According to the inspector, any processing step would negate the exemption from the PSM regulation.

We argued that the immediate dilution, making the ethanol non-flammable and non-hazardous, would not trigger the PSM standard. This was a huge issue for not only Heinz but also the whole vinegar industry. It would establish that companies making vinegar would come under the PSM standard, with a huge compliance cost. Heinz hired a law firm to represent the company. The case went to trial and fortunately Heinz won the case.

During my career I needed to terminate some employees. In one of the terminations, I was sued for wrongful discharge. In my opinion the employee had not made sufficient improvements in the department and therefore I decided that a change needed to be made. The employee sued, stating that he did nothing wrong. After sometime in litigation the suit was dismissed.

5

CONSOLIDATION OF MIDWEST GRADING AND BRINING INTO HOLLAND 1983

In 1983, Heinz had several tank yards in the Midwest. Midwest tank yards were located at the Holland factory and in Zeeland, MI; Lakeview, MI; Big Rapids, MI; Old Mission, MI; Plymouth, IN; and Fremont, OH.

The largest cucumber receiving and size grading operation was located in Lakeview, MI. These tank yards and size grading operations were installed many years ago. The tank yards included hundreds of 14 x 8 foot wood brining tanks with a capacity of 12,000 gallons each. Many of these tanks were installed in the 1920s.

Cypress tank yard Holland

At this time the cucumber and pickle management structure was organized into separate Agriculture and

Manufacturing groups. Both groups reported to separate management groups in the Pittsburgh headquarters. All cucumber grower contracting, receiving, size grading, and brining operations, except for the brining at the Holland factory, was managed by Carl Landis, Regional Manager of Agriculture. George Daily was the Holland Factory Manager responsible for Holland pickle manufacturing and the factory brining tank yard.

A restructuring project consolidated all of the Midwest brining and the cucumber grower receiving and size grading operations into the Holland factory. This was a huge project resulting in hundreds of grower truckloads of cucumbers being delivered into Holland daily during the summer green season from early July through mid-September.

The consolidation project started during the 1984 green season. This was an extremely difficult startup given the size and scope of the restructure project.

Cucumber grower receiving, unloading and size grading operations were delayed. Grower trucks were backing up into Holland streets. Growers were fuming because they needed to get their trucks back to the fields to pick up the next loads.

Charlie Roberts was the VP of the Packaged Goods Division and Jim Parker was the General Manager of Manufacturing responsible for the Holland factory. I am sure there was significant pressure from the Pittsburgh Agricultural group to deal with the consolidation issues at Holland. In response, Charlie and Jim scheduled a trip to Holland.

Everyone in Holland was aware of the consolidation issues. Most of us were also aware of the scheduled visit by the two men.

During the day of the visit, I got a call from Pam Petzak, George Daily's administrative assistant, which totally changed my career. I was told to stay in my office and to be in George Daily's office at 11:00am. I wondered if this would be my last day working for Heinz.

I went into George's office. This was the second encounter with Charlie Roberts at that point in my career.

I was told that Charlie and Jim were here to make some reorganizational changes to George's staff. They said that George was so upset about these changes, of which he had been unaware, that he resigned immediately and walked out of the factory.

They said that we want you to be the new Factory Manager effective immediately. I responded that I needed to go home and discuss this with my wife. In reality, I just wanted to get out of the office to have some time to digest what just happened. I had a good working relationship with George Daily.

I was told, "Okay, but you need to be back here within an hour because your new Production Manager and Financial Controller will be here at 1:00 and someone needs to show them the factory." I went home. Kathy told me that I need to make the decision and she would support me in whatever decision I make. Since there were not any advance announcements of the reorganizational changes, I felt compelled to accept the Factory Manager position. I returned to George's office and did so. I realized how disruptive this was going to be for everyone at Holland Factory and since most of the employees knew me, hopefully, I could reassure them.

Charlie told me that I had done a good job in my current technical services position. He further said that we would announce to Heinz that I would be the interim Factory Manager for six months. After six months I could make a decision to be the Factory Manager or go back to my prior technical services position. Charlie also said that he had asked Don Heckman, who was the Muscatine, IA Factory Manager, to come to Holland and train me to be a Factory Manager.

6

HOLLAND FACTORY & AGRICULTURE MANAGER 1984

Larry Heckel and Ed Balint arrived at the factory around 1:00. We had never met each other before that day. Larry Heckel was the Meat Products Module Manager at the Pittsburgh factory. Ed Balint was in the Pittsburgh headquarters Finance and Accounting Group.

I went to work that morning reporting to Jim Flynn as Manager of Technical Services for Pickles, Vinegar, and Sauces not the interim Holland Factory Manager. Larry and Ed came to the Holland factory expecting to be reporting to George Daily not Jerry Shoup. This unusual situation forged a bond between us to do everything necessary to keep the factory from failing for the next 28 years.

Because of my prior two years at the Holland factory as Manager of Technical Services, I knew almost everyone at the Holland factory. This was an awkward situation for everyone involved. No one in the factory was able to anticipate what was happening. We already had a Production Manager. The Factory Controller had recently retired and therefore there was an opening in this position. We walked around the factory and I introduced Larry and Ed.

The Production Manager told me he felt like a piece of fine china which just fell off the shelf and broke into pieces. Within a few days he left Heinz and went to Vlasic's factory in Imlay City, MI.

Don Heckman made a reservation in the Wickwood Inn bed and breakfast in Saugatuck. He called me and told me to meet him the next morning for breakfast in Saugatuck. I thought this was strange because he was there to teach me to be a factory manager and it seemed to me we should be meeting at the Holland factory.

At breakfast I asked Don what is your plan for the day? He said, "We are going fishing." He had chartered a fishing boat for the day. He told me, "Your first lesson is you need to learn that you do not need to be at the factory." He said, "You need to ensure that you have the right staff managers who are highly competent so that you do not need to be there." A few days later he left, saying you already know what you need to know.

The major issue facing our new team was the cucumber grading and grower receiving operations. This was the issue which had forced the recent reorganizational changes. Larry Heckel, as Production Manager, took charge of the situation. I remember Larry's frustration with the mud separation pit. This was a truck unloading step which was designed to separate field soil coming in with farmer loads but failed to work as designed. The pit would fill up with mud, requiring a major effort to empty the pit every day. Thanks to efforts by Larry and his Process Control team, the receiving operation was improved but ultimately needed to be eliminated. At least the Agriculture group did not complain as vociferously as they did previously.

I now reported to Jim Parker, General Manager of Manufacturing in Pittsburgh. Jim did not know me and I am sure he was concerned that I was not the best person to be factory manager. He asked Frank Adamson, who was a statistical efficiency expert, to move to Holland to advise and assist me in managing the factory and to conduct studies to improve productivity. Frank was in Holland for only a short period of time before being reassigned.

We needed to fill some critical positions to strengthen our management team in Holland. I traveled to the Pittsburgh headquarters and learned about Joe Kuzmanko, who was in the Pittsburgh factory. I talked to Joe. He told me he was ready for a change. I was very impressed with Joe's passion and can-do attitude.

Joe was eager to come to Holland. He became the Department Head of Facilities Maintenance. He eventually took charge of Production Maintenance. Everyone in the union and management groups respected and liked Joe, given his passion to improve productivity. Joe was a key member of our management team and a driving force pushing the production lines.

For years we depended upon Lester Perkins who was our key production mechanic on our critical vinegar lines one and two to make those lines run at maximum efficiency.

Ryan Hunderman supervised Facilities Maintenance. I always felt that Ryan and the facilities maintenance group did an outstanding job maintaining and keeping the factory running. Every time there was a snowstorm or blizzard forcing a factory shutdown, I knew that Ryan and the facilities maintenance group would be at the factory to handle any issues and make sure the next startup would go smoothly.

We also needed a Manager of Engineering. Our Manager of Engineering had been transferred to the Pittsburgh factory to help restructure that facility. I started a search for an outstanding Engineering Manager.

I was concerned about some of the corporate engineers and some of the engineers who had previously worked at Holland. Too often, their attitude was "I am the engineer and I know what is best for you. You need to accept my installation and make it work."

At the time we had major construction projects pending to replace our deteriorating wood brining tank yard and our desalting salt house operation. These were critical, must do projects to keep the Holland factory viable.

We were fortunate to hire Bill Snyder. Bill was the Manager of Engineering for Green Giant. From his resume and discussions with references, I learned that he was very successful with major projects and had a customer service attitude. His engineering approach was, "How can I help you and what do you need." This attitude was a breath of fresh air for me.

The D section of our wood brining tank yard and the old salt house had deteriorated over several decades. Replacement was absolutely necessary for safety reasons and to maintain operations. These were huge projects over successive years. They involved converting all of the old wood tanks to new fiberglass tanks and building a new Process Control building to efficiently desalt pickles.

Fortunately Bill came on board to engineer these projects. The cost for these two projects was approximately $3 million. This was a huge investment for the pickle business given that pickle market share and sales were declining. As we explained to upper management, we did not have a choice. We needed to spend the capital to safely maintain operations. The capital appropriations were approved.

For the next two years, I had several sleepless nights just thinking about the monumental task of removing hundreds of old wood tanks and building a new Process Control Building for de-salting pickles. While the construction was ongoing, we were contracting with growers for the upcoming crops. Our Heinz pickle business depended upon these projects being completed on time. Bill kept telling me to trust him, that he had it under control and would finish on time and on budget and that is what he did.

New fiberglas tank installation

The wood tanks we replaced were erected in the 1920s. We decided to preserve the history of the wood tanks by using the tank staves to erect a security fence around the 28 acres of Heinz property.

I remember many times going by the Heinz factory on my boat and being embarrassed by the black smoke emanating from our old coal boiler. At one time we believed the way to save money was to burn more coal. Bill was a key driver in converting to natural gas by decommissioning the coal boiler and installing a new high efficiency fire tube boiler. Not only did this reduce cost but also, after the conversion, I finally

Wood tank stave fence

enjoyed taking a boat ride by Heinz and seeing clear blue sky.

Bill became very interested in making vinegar and the vinegar operations. He worked closely with vinegar equipment generation companies like Frings and other suppliers to identify opportunities to modernize and optimize vinegar generation. His vinegar acctator installation projects significantly increased our vinegar generation output.

Bill designed a sleeving machine for Heinz Picnic Pack varieties which included squeeze bottles of ketchup, mustard and relish as single bottles sleeved together in a picnic pack sleeve.

Bill gained respect for his engineering skills from not only the Holland Team but also Pittsburgh headquarters. After his retirement he was called back for vinegar projects at the Heinz Fremont, OH factory.

After Bill retired, Terry Prins was promoted to Manager of Engineering. Terry had previously reported to Bill as a project engineer. Terry was also the Holland Factory Licensed Waste Treatment Operator. This was a critical position mandated by the State of Michigan.

Marena Rash was the project engineer reporting to Terry. With Terry's promotion, Marena needed to become the new State Licensed Operator.

As Manager of Engineering, Terry managed a huge project to install an automated Picnic Pack Collation production line. We were very proud to have the Heinz Picnic Pack production line located in the Holland factory. Sales for Heinz Picnic Packs have grown dramatically and now are sold in the U.S. and Canada.

The Holland factory was fortunate to have Chuck Eberly move to Holland as the Warehouse Manager. The warehouse was always well organized. For many years visitors to the Holland factory commented about how clean and well organized our warehouse was being maintained.

Larry Heckel was a strong, dominant production manager. He was always in charge and responsible for many improvements in the Holland factory. His certified operator concept, in which machine operators could earn a higher pay rate for learning how to make machine adjustments and perform minor maintenance on machines, was a real productivity opportunity. Unfortunately the union objected to the concept.

Larry and Kurt Schonfeld organized the daily production meetings, setting the proper tone and focus for each day. Kurt was the Assistant Production Manager and worked closely with the Pittsburgh production planning and inventory group. Kurt clearly understood potential problems in scheduling the various pickle, vinegar and sauce products. This understanding of the synergies in running certain product mixes increased productivity and optimized production labor utilization. Larry and I always thought the Holland factory rather than Pittsburgh personnel should have been responsible for working with sales and marketing to optimize production scheduling and inventory control.

Ed Balint had a reputation in Pittsburgh as being among the best factory controllers at Heinz. His monthly forecasts were right on target. This made my life easier. There is nothing worse than missing your financial forecast for the month.

The forecast may miss the budget standard set up at the beginning of the year for many different reasons, including poor sales projections and variable production.

The key is to know about the missed forecast early on, so that you can apprise headquarters. I had great respect for Ed's financial budgets and forecasts.

Ray Abney was an assistant controller working for Ed. Ray was very knowledgeable concerning all aspects of the factory operating budget. He was a key person developing the annual operating budget and standard costs. He was a go-to person for department managers to get updates and answer questions concerning department budgets.

When Ray Abney retired we eventually hired a new assistant controller, Doug Vaughn, reporting to Ed Balint. Doug was an excellent assistant controller who performed well. When Ed Balint retired Doug was promoted into the factory controller position reporting to me. Doug's performance was strong and I appreciated his forecasting and budgeting skills.

Jill Emmick and Gerard Maat worked for Ed and managed the payroll department. I had great respect for both of them. They were key members of the management team, resolving pay issues for all bargaining unit employees. Gerard was also well known in the community for being a foster parent for many young people for many years.

In many ways Mike Mooney, Quality Assurance Manager, and Evan Dawdy, Assistant Quality Assurance Manager, were my eyes and ears concerning the quality of our products being shipped. Mike and Evan were quick to alert Larry Heckel and production supervisors concerning any issues. I could always trust their judgment concerning the proper disposition of a quality hold.

Mike also took the initiative to make sure the property surrounding the factory was clean, neat and well maintained. He scheduled the lawn and landscaping care companies so that the grounds always looked good. Mike took the initiative to erect metal tulips, which were painted and attached to the inside of the pickle tank wood fence. The metal tulips appeared to be peeking over the fence along the Heinz Waterfront Walkway during Holland's annual Tulip Time festival.

Wood tank fence with tulips

Mike was my successor to Pickle Packers Int. and Evan was my successor to the Vinegar Institute. I always heard positive comments about both of them from PPI and the VI later in my career.

The Holland factory experienced considerable turnover in Human Resource Managers. Fortunately, Shannon Chada became the HR Manager. Shannon was a strong HR manager. I needed her advice and support during some major reorganizations, which were required to keep the factory viable. Olga Flores was also a key member of the human resources department. Olga was bilingual and served the department well. She was my translator from English to Spanish for many of my presentations.

A reorganization occurred at Heinz Holland, consolidating Heinz Holland Agriculture operations into the Holland factory. Dale Weigel, Manager of Holland Agriculture; Phil Wesorick, Holland Agriculture Financial Manager; and Anna Pavilon, Holland Agriculture Administrative Assistant, moved from the Agriculture Administration Building to the Holland Factory Main Office Complex.

Dale Weigel started to report to me as the Holland Agriculture Manager. Dale continued to be responsible for all cucumber grower contracting, cucumber purchasing from brokers and briners, apple and apple byproduct purchasing for cider vinegar, and all other agriculture produce purchasing, including cabbage and peppers.

Phil Wesorick was responsible for developing the cucumber standard costs and accounting for grower contracts and payments and all other agriculture produce purchases and accounting.

Anna accounted for cucumber and brined cucumber salt stock shipments and inventories. She also managed grower contract deliveries and payments.

I enjoyed working with Dale, Phil and Anna. They were dedicated team members who performed well in all of their agricultural responsibilities.

With the consolidation of Holland Agriculture into the Factory Operations, I assumed the position of Holland Factory & Agriculture Manager. The Holland Management Team was finally organized and remained intact through critical years for the Holland factory.

I was very proud of the Holland Management Team. The team worked very well together and dealt with serious issues as the pickle business declined, taking advantage of every opportunity to maintain the viability of the factory.

I was also very proud of the RWDSU local 705 team. Wanda Lacombe, Stella Soto, and Corey Ross were key members of the Bargaining Unit Committee and key members of the Maintenance and Production Teams. Even though we were on opposite sides of the negotiation table, we came together as one team to work together for the benefit of the Heinz Company and the Holland factory. Fortunately, we had a solid union/management team, which was well positioned to deal with the troubled times over the years ahead.

The team was keenly aware that 24 oz Kosher Dill Grocery Spears and #10 Kosher Dill Foodservice Spears were two large volume pickle varieties which were critical for our pickle business. The #10 spear variety was a premium product which commanded a premium price in the foodservice business.

For 24 oz grocery spears we needed to feed minimum five inch 3B size cucumbers to the spear packer so that the spear packing machine could cut off the top of the cucumber resulting in a flat surface. After cutting off the top, the spears would have the proper length to be place packed in the jar, so that the spears would fill the jar from the bottom heel to the neck of the jar.

For #10 spears, the Heinz quality standard specified a minimum of 3 ¾ inch length spears. Length of spears was important to ensure the spear could fit across a sandwich. 2B size cucumbers were required to give the proper length and diameter size to minimize short spears and to be able to quarter cut the cucumbers and meet the specified count range for the #10 can. In order to meet the product quality standards for these products, a significant number of inspectors was required on both the grocery and foodservice spear production lines to manually inspect and remove short or broken cucumbers.

The team identified a significant productivity improvement opportunity when we introduced a length grading machine. This machine was designed with an adjustable gap between two belts. Based on the length and center of gravity of cucumbers, short or broken cucumbers would fall through the gap, being rejected. Cucumbers with the right length would go across the gap and be accepted. For 24 oz spears requiring a five inch length, the gap was set at 2 ½ inches. For #10 spears with a minimum length of 3 ¾, the gap was set at two inches to accept four inch cucumbers and reject those under four inches. These length graders resulted in a huge productivity labor savings.

We realized significant cucumber yield cost savings with the introduction of water knife cutting for foodservice spears. Heinz owned the Ore Ida potato processing company. We learned that Ore Ida was cutting potatoes into French fries with a water knife cutting system, so we went to Ore Ida's factory in Greenville, MI.

The system pumped potatoes at high speed through a pipe which had a set of knife blades inside it. The knives cut the potatoes into French fries.

Prior to water knife cutting we utilized Robbins Cutters, which fed cucumbers through a series of accelerating belts that thrust the cucumbers through a set of knives. The yield loss was high; it was difficult to control the cucumbers entering the knives, so we had a high percentage of miss cuts.

We installed water knives. But, because cucumbers are shaped different from potatoes, we made improvements to the Ore Ida system. This gave us better control of the cucumber entering the water knives. These improvements allowed us to increase the line speed and cases per shift, lowering the cost of foodservice spears by increasing cucumber yield in terms of cases per bushel and productivity.

7

100-YEAR ANNIVERSARY OF HEINZ IN HOLLAND 1997

The H. J. Heinz Company was established in 1869. The founder, Henry John Heinz, first product was horseradish. Mr. Heinz was well known for his marketing expertise. Note the picture of his horseradish jar, circa 1871. The label shows Heinz Strictly Pure with a picture of a horse and the word "Radish" under it. The Heinz Company started packing pickles in 1871.

Henry John Heinz

We knew that Heinz had a rich history in Holland. As we approached the 100-year anniversary of the Heinz Holland Factory in June 1997, we asked Randy Vande Water, a well-known Holland historian, to write a book on the 100-year history of Heinz in Holland. Randy, as the former editor of the

Horseradish jar, circa 1871

Holland Sentinel, had access to every Heinz newspaper article from the start of discussions about Heinz coming to Holland in December 1896.

In 1896, Mr. Heinz wanted to find a location in Michigan to start a pickle business because Michigan was the largest cucumber growing state and still is today. Mr. Heinz needed a location in Michigan with a rail siding and boat dock so that he could ship his 50 gallon barrels of pickles and other packages throughout the Midwest. In December 1896, the Holland Mayor and Town Council were told that Heinz would start a pickle business in Holland, if the town gave Heinz two acres of land with a boat dock and rail siding and if local farmers and gardeners would commit to growing 300 acres of cucumbers for Heinz.

Holland Sweet Gherkin jar circa 1900s

The City purchased two acres of land with a rail siding and boat dock for $800 on the present day Heinz site.

Numerous town Hall meetings were held, soliciting acres of cucumbers for Heinz. By February 1897, 500 acres were committed. Heinz then notified the City of Holland that it would start the business. The original pickle building cost $2,100 and was commissioned in June, 1897.

To commemorate the 100 year anniversary of Heinz in Holland, we convened a dedication ceremony in Kollen Park on May 31, 1997. We also ordered special anniversary coins struck with a Heinz Holland logo designed for the 100-year anniversary. These words were on the obverse of the coin: "Dedicated to All Past and Present Employees Whose Work Ethics and Commitment Have Resulted in Prosperity and Growth throughout the 20th Century."

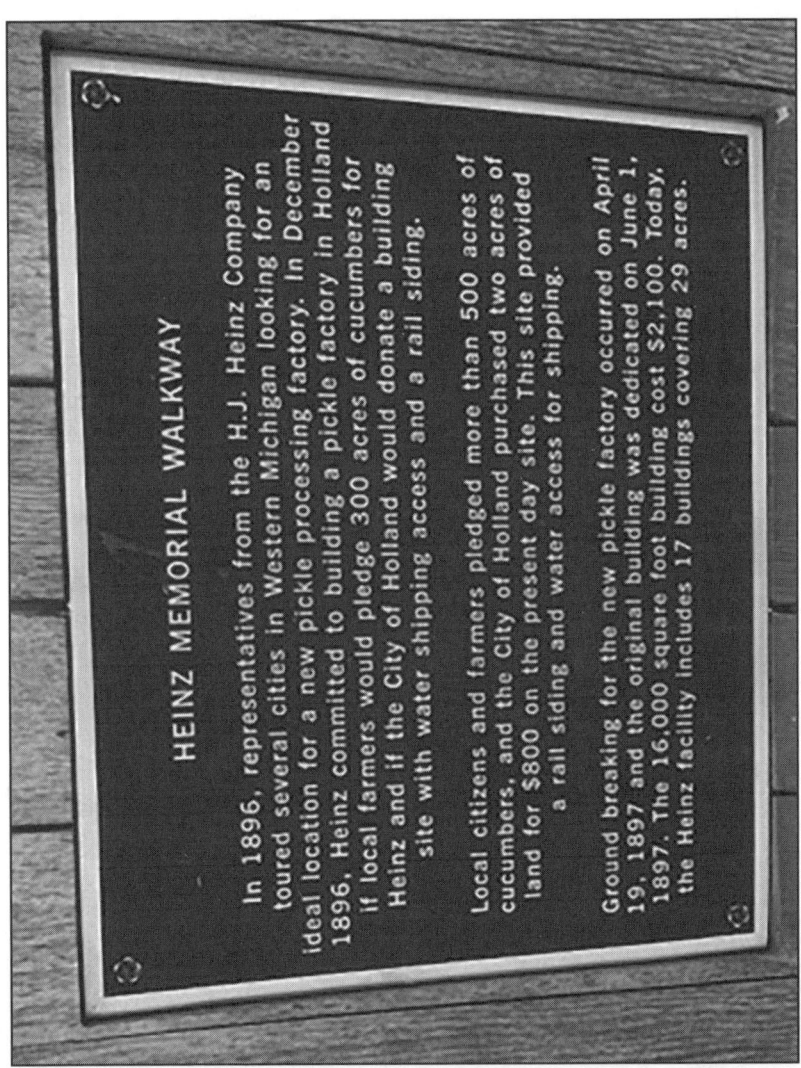

Heinz history Walkway plaque

First Factory June 1897

Heinz Holland Employees 1906

100-YEAR ANNIVERSARY OF HEINZ IN HOLLAND 73

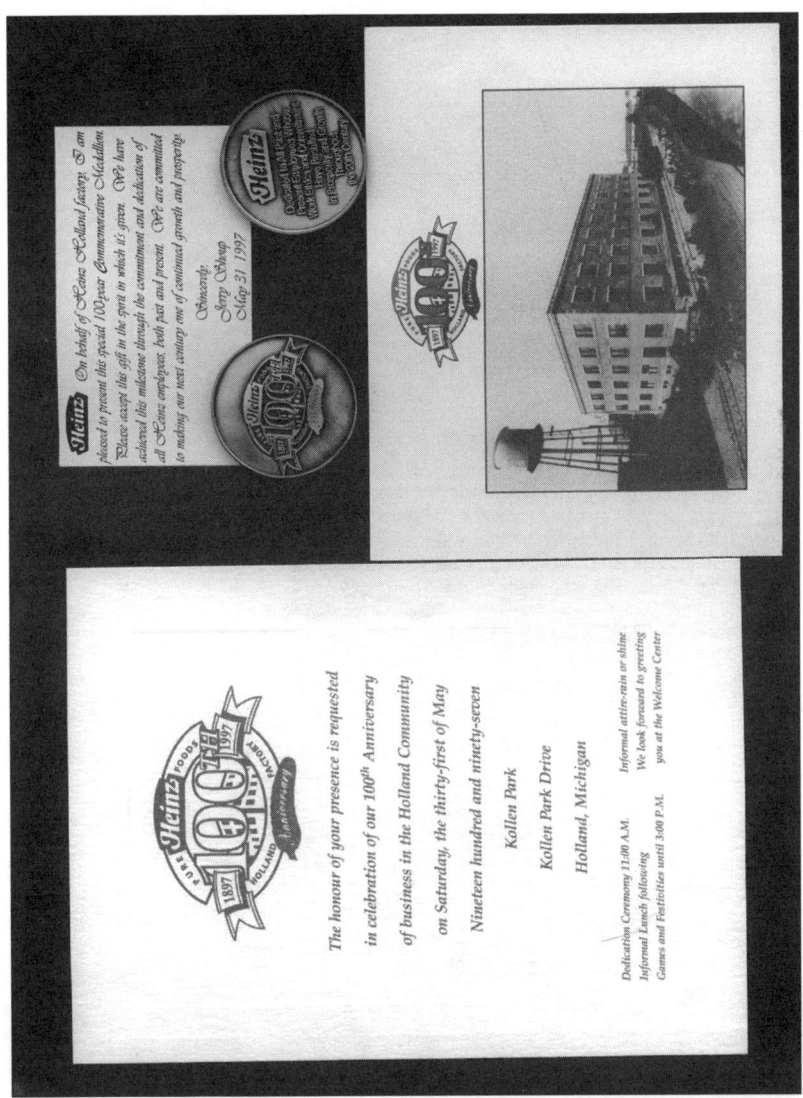

100 year celebration invitation

HOUSE OF REPRESENTATIVES
WASHINGTON, D. C. 20515

PETE HOEKSTRA
SECOND DISTRICT
MICHIGAN

May 31, 1997

H.J. Heinz Company
Holland Factory
Holland, Michigan 49423

Dear Friends :

I would like to extend my sincere congratulations on your Centennial Anniversary celebrating your commitment and participation in the Holland Community.

From the beginning when 200 farmers pledged 500 acres and funds to build the Heinz Factory, to the successful and expansive facility we see today, there is no doubt that the Heinz Company and the City of Holland have developed a close and prosperous relationship. This relationship has played a vital role in the success of this community.

As the city of Holland celebrates its Sesquicentennial, Heinz can take special pride in its contributions which have helped this thriving community become an "All American City." I am proud to join with state and local officials, as well as the entire Holland Community, in congratulating you on this momentous occasion.

Sincerely,

Pete Hoekstra
Member of Congress

PH/pvd

100 year anniversary letter Congressman Hoekstra

8

ADVERSITY IS A GREAT MOTIVATOR
1988

A few years later, a crisis situation developed as we prepared our new budget for the fiscal year beginning May 1. For budgeting each year, we were given a production schedule developed by the Pittsburgh headquarters' finance and accounting groups, based upon input from the sales and marketing group. We learned that the budget production schedule included 300,000 cases of #10 pickles, which were scheduled to cover potential salvage sales.

Our Packaged Goods Vice President stated we do not budget for salvage sales, therefore 300,000 cases of #10 pickles were removed from our production schedule for budgeting purposes. This created a million dollar hole in our budget target for that fiscal year.

Salvage sales occur when sales are soft and some finished goods inventory reaches the "best if used by" date. At the time, factory overhead for budgets was absorbed based upon the weight of the budget schedule case. #10 pickle cases were among the heaviest cases and absorbed a significant amount of fixed overhead.

Our Vice President was an unforgiving finance guy and was scheduled to be in Holland in a few weeks to review our budget. I once heard this Vice President say "Adversity is a great motivator". That comment stuck with me during my career. I used it at different times as a rallying cry to deal with difficult situations. The adversity created by this million dollar hole in our budget motivated our Staff to launch a project called Max Up which was a project to maximize human resources. We knew, to survive as a factory, we had to come up with a plan to deal with this adverse situation.

We decided to take our management group, including the factory staff and department heads, offsite for two days of meetings. We would use the time to brainstorm a plan to offset the million dollar hole in our budget. We rented a large facility in Saugatuck for the brainstorming meetings. To prepare for the meetings, we asked all team members to meet with the accounting department and brainstorm their departmental spending. We then asked them to come up with restructuring ideas to reduce headcount and develop other cost reduction ideas. At the time we had a total hourly headcount of approximately 285 employees. Over two days, each team member had to report a restructuring plan to reduce cost.

It became clear that we needed to focus on our production schedule. At the time, we were highly seasonal with 60% of our production schedule occurring in the fresh pack green season from mid-June, including the pre-season prefill, to mid-September.

For the eight months after green season, we experienced 16 weeks of shutdown on a double shift scenario. What if we went to a single shift scenario during the off-season? We could complete the production schedule on a single shift reducing 16 shutdown weeks to two.

This change would mean a reduction of 50 to 60 shift workers required to complete the schedule during the off-season. Additional cost reductions would be realized in wage rate differences when we hired more seasonal temporary workers at a lower wage rate at the startup of the double shift scenario for green season.

Thanks to the departmental restructures brainstormed during the meetings coupled with the single shift reduction in required shift workers, we were able to identify a total workforce reduction of 75.

We knew that we could complete our required annual production schedule with a reduction of 75 hourly workers from 285 to 210. We also identified department restructures, consolidating departments which resulted in a reduction of salaried management headcount. It was critical that the headcount reduction include both hourly and salaried management employees.

We realized that we would have increased cost from the single shift scenario due to the doubling of sanitation shifts as compared to the double shift scenario. Also we would have increased energy costs due to additional days in which we would need to run the boiler to generate steam.

The savings in benefit costs from the reduced head count coupled with the wage rate differentials realized by hiring more seasonal workers more than offset the cost increases from the single shift scenario. The bottom line was that Max Up covered the million dollar problem.

When our Vice President arrived for our budget review, he said you have a major budget problem here in Holland. We responded by saying, "Yes and here is our strategy to fill the hole." He was impressed and said he would fully support this strategy. I do believe that without this strategy Holland factory would have eventually been closed.

We wanted to treat all employees with dignity and respect. The loss of jobs was absolutely necessary to maintain the viability of the factory. Our focus was on making as many of the reductions in Max Up voluntary in order to minimize as much as possible the negative impact. Shannon Chada, HR manager, identified those hourly employees with only a few years of service and those with over 30 years of service. Shannon was aware of long term employees who were considering retirement and employees who would most likely opt for an enhanced early retirement offer. She also gave her opinion concerning how many employees with only a few years of service would consider resigning for a $2,000 payment.

Shannon worked closely with Chauncey Smith at the Pittsburgh headquarters human resources group to make sure that we were treating all employees fairly. We were successful in gaining support for enhanced early retirement packages and severance payments for hourly employees with only a few years service. We also developed severance packages for salaried employees affected by consolidations.

We met with the bargaining unit committee and explained why Max Up was necessary given our budget problem. We explained our plan to offer $2,000 payments to hourly employees with only a few years service who volunteered to resign and take the payment. We also explained the offer for voluntary enhanced retirement packages for long term hourly employees. We also acknowledged that some salaried management employees were being given severance packages who were affected by departmental consolidations.

We emphasized that bargaining unit employees leaving were leaving voluntarily. We emphasized that Max Up would control our factory cost and make our factory more viable for the future. We also emphasized that remaining employees will be laid off for only two weeks rather than 16 weeks. These were positive outcomes of Max Up. The bargaining unit committee, while not enthused about Max Up given the loss of members, did not resist our implementation of Max Up.

We realized that with Max Up we would lose many machine operators who were fully trained to gear up for the fresh pack green season. This would require training many new seasonal workers every year. We developed a program for hiring seasonal workers before the season started, to train them to run critical machines. This was not totally successful. We suffered from poor productivity for a few years given the lack of trained machine operators.

9

PERRIGO 1991

Our pickle business continued to decline after Max Up. We jumped on an opportunity which came our way during this troubling financial time. Dave Willard was a former Heinz Holland Factory Manager who left Heinz and eventually went to work for Perrigo, a drug manufacturer in Allegan, MI. Dave called me in 1991 with an opportunity.

Dave mentioned to me that Perrigo needed a co-packer to pack its 32 oz generic mouthwash for 2 to 3 years for 250,000 to 300,000 cases per year while Perrigo built a new factory in North Carolina.

Dave told me that he believed that Heinz Holland production line two, which is our vinegar bottling line, could fill Perrigo's 32 oz mouthwash. He said that he only needed to have the bottles filled, capped, case packed and palletized. The pallets would be loaded directly onto trucks and shipped to a facility, which would label the bottles for Perrigo. This sounded to me like a huge income opportunity given we had line time to fill the mouthwash.

Dave explained that mouthwash is an OTC (over the counter) drug and we would need to register with the FDA as a drug manufacturer. He also said that engineers from Perrigo would need to inspect and qualify our line two to ensure there was no spark risk, given that mouthwash has a low flash point; it contains over 20% alcohol.

We discussed the filling procedure. The water needed to be purified. Perrigo would prepare the mouthwash and pump it into trucks for delivery to Holland. Holland needed to install a separate fill line from the truck unloading point directly to the filler so that there was not any possibility of contamination from any Heinz product or city water.

We calculated our costs, which were minimal because there were not any pasteurization or processing steps. We simply would fill directly into Perrigo 32oz plastic bottles and then immediately cap and convey them, unlabeled, to the case packer and palletizer. We then would load the bottles directly onto trucks for shipment to Perrigo's facility for labeling.

With the help of the Pittsburgh headquarters, the registration was filled out and sent to the FDA. My phone number was given as the contact number for Heinz. Sometime later I was in my office and received a call from the Detroit FDA office. I was asked if we knew what we were doing. He asked me to explain. I went through the proposed process steps and the FDA said "Oh you are a re-packer, not a drug manufacturer." He said that we would still need to be inspected.

32oz Mouthwash bottle

We included a $1.00 toll fee profit above our actual cost to produce a case of mouthwash. This generated $250,000 income per year for two years, which helped offset ongoing pickle volume declines. Packing mouthwash for Perrigo was one of the key opportunities for our hang-on strategy to keep the factory from closing until other new products could come into Holland.

A few months after we filled our last case, the FDA finally arrived for the inspection. I told them we just completed our contract and so they left the factory. No doubt, the Heinz Holland drug inspection was low on their priority list once they found out we were a drug re-packer rather than drug manufacturer. Heinz Holland may have been the only Heinz facility in the 144-year history of the Company to register as a drug manufacturer.

10

CO-PACK OF COMPETITOR PICKLES 1995

In contracting cucumbers from growers, you must maintain a pickle business which includes all sizes of cucumbers. When a grower harvests his cucumber crop there are all sizes of cucumbers on the vine. This means to take a grower's crop you must have a sales mix of all the various pickle varieties which utilize the various sizes.

Sweet pickles utilize small #1 size cucumbers. Grocery bread 'n butter or hamburger slices utilize medium size #2 cucumbers. Foodservice hamburger slices and grocery spears utilize larger #3 size cucumbers. Relish utilizes large size cucumbers as well as misshapen or broken cucumbers.

As Heinz grocery pickle sales continued to decline our team faced a real dilemma because this created an imbalance in our crop procurement size requirements. We needed to take steps to better utilize our incoming crop.

As mentioned in a previous chapter, our grocery spear production line with its Solbern spear packing machines, gave us an opportunity to co-pack grocery spears for other pickle companies. By attending Pickle Packer meetings, we had the opportunity to meet with other pickle companies and find companies who needed a co-packer for grocery spears.

For some years, we were able to co-pack 24 oz grocery spears for Steinfeld Pickle Company, Nalley Fine Foods, and Paramount Pickle Company. We were able to develop co-pack prices which generated significant income. This was another opportunity to hang-on and keep the factory doors open. Co-packing for competitors also helped our crop procurement by balancing our crop size requirements and utilization.

At this time, we were trying to keep the Holland factory doors open by generating income from co-packing competitor pickles. It is ironic that at the same time Heinz Pittsburgh was making a concerted effort to find pickle packers to co-pack Heinz Pickles to close Holland factory doors.

If Pittsburgh could find a co-packer cost which made sense for the Heinz pickle business, then it would be feasible to close the Holland factory by transferring all other Holland products to other Heinz factories. Fortunately Heinz Holland costs were being controlled and were lower than any co-pack bid.

I do not believe Heinz wanted to exit the pickle business. Pickles were the founder's favorite product. At the 1893 World's Fair, Mr. Heinz introduced the Heinz Pickle Pin and gave away

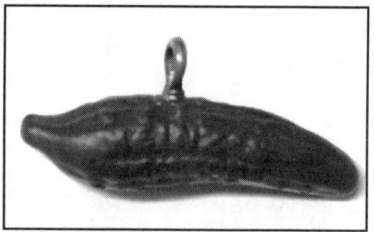

Heinz Original 1893 Pickle Pin

one million pickle pins. The Saturday Evening Post called the Heinz Pickle Pin "one of the most famous giveaways in merchandising history."

Many consumers still think of Heinz as a pickle company. A Heinz pickle is on the label of many Heinz products. I remember Heinz contacted me every year before the Heinz annual stockholders' meeting and asked what was new in the pickle business. At every annual meeting there would be a perennial question about Heinz pickles. By co-packing pickles Heinz could remain in the pickle business and close the Holland factory doors.

PROJECT MILLENNIA LATE 1990S

Tony O'Rielly was Heinz CEO during the 1980s and 1990s. His performance as CEO was outstanding. Heinz was known as the darling of Wall Street. Heinz stock price increased significantly during his tenure. I can remember at least two stock splits.

In the late 1990s Tony introduced Project Millennia to restructure the Heinz Company for future growth in the new millennium. He stated that one of the five major factories in the U.S. would be closed. At the time, the five major factories included Pittsburgh, Fremont, Holland, Muscatine, and Tracy. The *Holland Sentinel* was very interested in the press article and started to speculate as to which major factory would be closed.

Unfortunately, Tony O'Reilly granted an interview to the Associated Press concerning Project Millennia. The AP asked him what his biggest mistake was as CEO of Heinz. He stated "staying in the pickle business". As Bernie Sanders would say that statement was "HUGE" as it impacted all Heinz Holland employees, including me. The *Holland*

Sentinel speculated that Holland would be the major factory closed based upon Tony's comment to the AP about the pickle business.

This started a rumor in Holland that the Holland factory was closing. This greatly affected all Heinz Holland employees, including me.

I believed in giving at least quarterly updates to all Heinz Holland employees on all shifts concerning factory performance, covering productivity, quality and safety. I also included in the updates an evaluation of how the Heinz businesses served by our factory were doing and whether or not there were any new product initiatives. Employees came to me, some with tears in their eyes, asking why I did not tell them the factory was closing.

All factory managers participated in weekly phone conferences with the Pittsburgh communications department and other headquarters management personnel to give updates on their factories and updates concerning the progress of Millennia.

I knew that the factory to be closed was Tracy, not Holland. All factory managers were warned not to make any comments concerning factory closures or any statements about Project Millennia. We were instructed to refer all questions to the Pittsburgh communications department.

To add fuel to the fire, the *Holland Sentinel* stationed reporters just outside the factory gate to interview employees heading to the parking lot concerning their feeling about the factory closure. The only comments I remember seeing in the paper were "...we like management here..." or "...we are like a family here."

The *Sentinel* kept calling me at home. I told Kathy to tell the reporter that I was in the shower. They kept calling.

Finally the reporter said he must be the cleanest guy in town since he is always in the shower.

I remember being in a daily production meeting and Larry, our production manager, asked why don't we give the *Sentinel* reporter a real story to report. He said, "Let's get a mannequin, dress him up like a maintenance mechanic with a yellow hard hat, take him to the 3rd floor roof and throw him off the building in front of the reporter." A little humor at this time went a long way.

An intolerable situation developed when we learned that some of our employees were being turned down for loans because the factory was being closed. We discussed this with Pittsburgh communications and said something needed to be done to address the situation. Pittsburgh communications agreed and said they would contact the *Holland Sentinel* and tell the *Sentinel* that Heinz considered pickles to be a core Heinz product. This ended the rumor.

12

TRANSFER OF VINEGAR BOTTLING AND MUSTARD, WORCHESTERSHIRE, HORSERADISH, AND JACK DANIELS BARBEQUE SAUCES INTO HOLLAND FACTORY 1999

We could sense that Heinz Holland was starting to get a lot of credibility and respect for steps we had taken to control our costs despite declining production volume. We started to challenge upper management as to why white distilled vinegar generated in Holland was being shipped to the Pittsburgh factory for bottling and then shipped to customers, when it could be bottled in Holland and shipped directly to customers. This challenge ruffled many feathers in Pittsburgh.

At this time, I reported to Dan Nolan, General Manager of Manufacturing, at the Pittsburgh headquarters. I reported to Dan for 12 years starting in 1994. During my 38 years at

Heinz I worked for several bosses. I had more respect for Dan than any other boss in my career at Heinz. He was smart with good business sense. He would carefully listen to all factory managers and was able to quickly analyze an issue and make the right decision. He was very supportive when it was appropriate to be.

Dan Nolan

I am sure Dan heard the challenge concerning shipping vinegar from Holland to Pittsburgh for bottling, and I know he worked with the headquarters' financial and distribution managers to do the right thing for Heinz. I remember getting a call from the Pittsburgh Factory Manager complaining about my trying to steal his vinegar. I thought to myself, "Welcome to my world, trying to deal with lost production volume."

In February 1999, Heinz announced the Operation Excel Restructuring Project, which included cutting about 4,000 jobs over the following four years and closing or selling 20 factories worldwide. As part of Excel, the Pittsburgh factory was restructured to become a soup and baby food factory. As a result, vinegar, worcestershire sauce, mustard and horseradish sauce were transferred from Pittsburgh to the Holland factory. Shortly thereafter, all remaining vinegar bottling in Heinz USA, at the Muscatine and Tracy factories, was also transferred to the Holland factory. A few years later, Jack Daniels barbeque sauce was transferred into Heinz Holland from the Heinz Fremont factory.

Finally our Holland team strategy of hanging-on by controlling costs through troubling financial times was successful. All the new production stabilized our production volume and increased the profitability of our product portfolio.

SWANSON PICKLE COMPANY CONTRACT 2000

As discussed in chapter five, with the consolidation of all Michigan grower grading and brining operations into the Holland factory, there were major issues with the grower receiving and grading operation. Richard Geiger was the head of the process control department. The process control department managed the grower grading and brining operations. Process Control worked with Facilities Maintenance and Engineering to make improvements that allowed the grower grading and receiving system to operate for several years after the consolidation in 1984.

In 1999, the old system needed to be replaced for safety and operational issues. The cost to replace the system would exceed $2 million. We knew that proposing to spend $2 million for the pickle business would not be a good career move. It became clear that the only option was to outsource the grower grading and receiving operation.

Dale Weigel, Manager of Agriculture, and I discussed approaching Swanson Pickle Company in Ravenna, MI to determine if the firm would have any interest in taking over Heinz grower grading and receiving operations and expanding brining for Heinz. Swanson Pickle Company sold cucumbers to Heinz and brined cucumbers for many years. We had the utmost respect for Don and John Swanson and knew Swanson Pickle Company would do an excellent job for Heinz.

John and Don Swanson

Dale and I went to Swanson Pickle Company. They agreed to negotiate a contract which they could take to the bank and secure the necessary loans.

The tank-brining project was for 500,000 bushels per year. The total investment for the grading and brining was $4.175 million. $3.175 million came through the Michigan Strategic Fund and $1 million came through a conventional note.

The grading contract was signed on September 22, 2000 followed by the brining agreement on November 15th. As mentioned in chapter six, length grading was critical for productivity for grocery and foodservice spears. We needed Swanson to handle length grading for us in order to maintain that productivity. We signed a seven-year equipment lease with Swanson for the length graders and other grading equipment. After the seven-year lease the grading equipment was purchased by Swanson for $1.00.

The grading contract with Swanson Pickle Company was critical to keep the Holland factory viable. It avoided an investment of $2 million in the declining pickle business and eliminated a significant safety issue for our employees.

It also eliminated up to 100 farmer trucks coming into Holland every day during the summer season with all the dirt and dust from the farmer fields. Eliminating this traffic congestion and making the Heinz factory and surrounding residential area adjacent to Kollen Park cleaner and more pleasant for all of us may have been the greatest impact of the Swanson contract.

AGREEMENT

THIS AGREEMENT, made and entered into this 22nd day of September, 2000, by and between the Heinz North America, Division of H. J. Heinz Company, a Pennsylvania corporation, having its principal place of business at 1062 Progress Street, Pittsburgh, PA 15212 ("Heinz"), and Swanson Grading and Brining, Inc., a Michigan corporation, having its principal place of business at 11561 Heights-Ravenna Road, P.O. Box 211, Ravenna, MI 49451-0211 ("Swanson").

WITNESSETH:

WHEREAS, Heinz desires to obtain grading, tanking, storage and/or transportation services of cucumbers ("Cucumbers") under the terms of this Agreement and Swanson desires to supply such services to Heinz under the terms of this Agreement; and

WHEREAS, in order to minimize the costs of performing the services contemplated herein Swanson requires the use of, and is willing to lease and / or buy certain equipment, as more fully described herein and Heinz desires to minimize the costs of purchasing the services contemplated herein and has such equipment and is willing to lease and / or sell such equipment to Swanson.

NOW, THEREFORE, in consideration of the premises, the mutual promises contained in this Agreement, and other good and valuable consideration, receipt of which the parties acknowledge, the parties, intending to be bound legally, agree as follows:

ARTICLE I: TERM OF THE AGREEMENT

1.   Term. The initial term of this Agreement shall commence as of the date first written above and shall continue in full force and effect until December 31, 2007, unless terminated earlier per Section 16 or 17 below ("Initial Term"). This Agreement may be renewed for a term of at least one additional year beginning on January 1, 2008 and for additional renewal terms thereafter. In order to renew this Agreement, both parties must agree in a writing to a renewal term by no later than the January 1st before the end of the initial term or the renewal term, or extend the deadline for such an agreement in a writing signed by both parties. In light of the planning required to transition the activities contemplated herein, the parties agree to have a formal in person meeting by no later than November 1, 2006 before the end of the term of this Agreement to discuss the terms of a possible renewal term.

Swanson grading agreement

14

VLASIC ACQUISITION PROJECT

Another Millennia initiative was to acquire #1 market share brands. During this time Vlasic Foods went up for sale. Vlasic Foods included Vlasic pickles, Open Pit barbeque sauce, a frozen food entrée company, and a British company. Heinz was interested in both Vlasic pickles and Open Pit barbeque sauce, both #1 market share brands.

Heinz made a bid for only Vlasic pickles and Open Pit barbeque sauce. Heinz was the only company to make a bid for any of the businesses and so the Heinz bid proceeded toward a sale.

Dan Nolan and I visited Vlasic pickle factories in Imlay City, MI and Millsboro, DE. We invited Vlasic Manufacturing and Agriculture Managers to Pittsburgh for discussions. We had an excellent plan for Heinz and Vlasic pickles.

During the due diligence period the Federal Trade Commission (FTC) interceded and blocked the sale claiming it would create a monopoly in the relish business. I was asked to go to Washington with the Heinz corporate attorneys as the pickle expert to plead our case.

The pickle business was very diverse and included many different small and large pickle companies. At the time, Heinz had approximately an 11% share and Vlasic had approximately a 27% share of the pickle business. So the combined Heinz and Vlasic Pickle business market share was less than 40%.

During the meeting, we told the FTC that their position was ridiculous because there is not a relish business separate from the pickle business. As we explained, the Pickle business includes relish, kosher dills, sweet pickles, hamburger dill slices, and other varieties.

We further explained that a relish variety is really a byproduct which supports the total pickle business because it uses various sizes of pickles which are broken, misshapen or crooked and cannot be utilized for other pickle varieties. This fact makes it not practical to think of relish as a separate business.

We scheduled a second meeting for the next day. That evening we went to several grocery stores and purchased many different pickle products including relish from Vlasic, Mt. Olive, and other companies. The next day we lined up all the different pickle varieties from the different companies to review with the FTC.

We told the FTC this is the pickle business for Vlasic and Mt. Olive. Relish is only a small part of the total business. The FTC continued to be unimpressed by our presentation.

Heinz appealed the FTC decision to a three-judge panel. The panel ruled two to one in favor of Heinz. I started to dream about Heinz crowning me the Pickle King. During Henry J. Heinz's lifetime he was known as the Pickle King.

The day before the Heinz acquisition of Vlasic was scheduled to close, Tom Hicks from Hicks Muse made a bid for all four Vlasic companies and as a result Vlasic Foods was sold to Hicks Muse and eventually became a part of Pinnacle Foods. I saw my one and only chance of becoming the next Pickle King go up in the air like a puff of smoke.

Heinz Holland City Cooperation

15

HOLLAND CITY AND HEINZ COOPERATION AND MUTUAL RESPECT 2007

In the early 1970s Heinz Holland drilled three deep wells approximately 6,000 feet deep into a salty porous rock layer for disposal of its food processing wastewater. The wastewater was injected under high pressure into a porous rock layer which absorbed and contained the wastewater.

The food processing wastewater was less than 1% salt and mildly acidic with a pH ranging from 3.5 to 4.0. The mild salt solution resulted from the brining and desalting operations associated with pickle processing. The mildly acidic wastewater was due to the brine fermentation resulting in lactic acid and also from vinegar generation, vinegar bottling and the use of vinegar in making the various Heinz products in Holland.

For years Harrell Daniels, Abundo Almanza and Brenda Byrne ran the waste treatment department. This was a critical function which affected our ability to maintain factory operations.

Whenever anyone in the community asked me about the deep wells, I responded truthfully that salt is mined from the earth and sea and we are returning it to the salty rock layer in the earth. Also our wastewater is a food processing wastewater which is not toxic.

In the early 2000s the capacity of the wells started to decline due to the age of the wells and slow plugging of the injection zones. The wells were costly to maintain and operate. Periodically we needed to pull the well pipe and replace 30 to 40 foot sections. We needed to plan for the future. If it were possible to get new permits to drill new wells it certainly would take months to years and cost millions. The capital cost for new wells would be prohibitive for the pickle business.

Since Heinz's founding in 1897, there has been cooperation and mutual respect between the Holland City and Heinz Holland. This certainly was true in 2006 and 2007 when the city approached Heinz about taking Heinz's wastewater. The city had the capacity to take the wastewater and testing of the wastewater proved it would be beneficial for the city, given its mildly acidic pH. The agreement between the city and Heinz to take the Heinz wastewater solved the deep well dilemma facing Heinz and offered a long-term solution.

Another key example of the City and Heinz working together to solve problems occurred some years ago. Heinz always shuts down for preventative maintenance over the holidays from mid-December to early January.

One year, a main Heinz transformer failed in mid-December. This transformer was absolutely necessary for

Heinz to re-start production in January. To order a replacement transformer would take many weeks and would be a major customer service disruption and loss of business for Heinz Holland. The BPW had a spare transformer which they gave to Heinz from their inventory with the understanding that Heinz would replace the transformer after an appropriation was approved and a new transformer was ordered and received. The startup was on time without any customer service issues. Thank you Holland City BPW.

16

CO-PACK OPPORTUNITIES INDIA AND MEXICO 2007

In 2006 and 2007, Heinz started to experience a significant reduction in the number of seasonal workers available for seasonal hiring. We realized that we would need to take steps to reduce the impact of the shortages.

At Pickle Packer meetings, we were introduced to Global Green Corporation, GGC, an Indian Corporation that was offering co-pack capability for labor intensive hand pack varieties like grocery spears and other pickle products. We decided to explore this opportunity.

I went to India on two separate 10-day trips. On the first trip, Larry Murray from the headquarters Quality Assurance group went with me. The main purpose of this trip was to review the manufacturing and agricultural operations in Hyderabad in order to preliminarily qualify GG as a co-packer for Heinz.

The flights from Detroit through Amsterdam to Hyderabad took 23 hours. We arrived around midnight Indian time. I must admit it was a shock when we de-planed, went through customs, and then into the ground transportation area. There was a sea of people, all waving signs, trying to find their visitors and guests.

Fortunately we traveled with Paul Palnikar who was a technical director for Vlasic in the U.S. who had recently left Vlasic and gone into consulting. Paul was an Indian and fluent in Hindi. Hindi is the primary language in India. He was a sales representative for GG.

GG purchased a tobacco processing plant and converted it to a food processing plant. We conducted factory batch trials for a few Heinz pickle products. The factory was given a preliminary qualification as a Heinz co-packer pending a review of the factory batch trials of the Heinz pickle products in Pittsburgh by the Quality Assurance and Marketing groups.

Final co-packer approval required a more in-depth agricultural operations review. We spent a few days in the cucumber grading and receiving facilities but knew that we needed to return to India in a few months when the harvesting of the new crop started.

I have vivid memories of driving on the roads in India. We had a chauffeur who would drive the GG executives, Larry and me from the hotel to the factory or to the agriculture grading stations. On the highways, you would drive on either side of the road wherever there was an opening, swerving to miss trucks directly in front of you. The drivers were incredibly skilled, like race drivers.

After the factory and preliminary agriculture reviews, it was time to go the GG corporate headquarters in Bangalor. I remember driving around Bangalor and seeing the high tech companies, including HP, IBM and others. I was also impressed seeing grade school children marching to their classes in blue and white uniforms. The GG executives said children were eager to go to school. Indian children were totally committed to going to school and getting advanced degrees as a way to get out of poverty. Perhaps I was only seeing the well-to-do children. I am not sure.

In Bangalor, we had business meetings. I learned that GG currently supplies pickles to dollar stores in the U.S. and large chains in Canada. I also learned GG could pack labor-intensive hand pack grocery spears and ship them halfway around the world to the U.S. and Canada in sea containers at a competitive cost.

The typical cost for an Indian factory worker in U.S. dollars was less than 20% of the cost of a U.S. worker. Cucumber costs were also much lower and if I remember correctly some of the cost was subsidized by the Indian government. I also learned that Indian pickles were subjected to a significant import duty into Canada.

In GG's mind this was a one-way street, they would co-pack for us. But I started to think this should be a two-way street. Heinz Holland could co-pack low labor pickles, such as slices, for GG and GG could co-pack high labor varieties, such as grocery spears, for Heinz Holland. I did not know how NAFTA might apply in this case, shipping GG pickles packed in Holland, MI into Ontario, Canada.

In my mind we should have been talking about a joint venture but GG only wanted to talk about a one-way street. I suspect that the Indian government subsidies may only apply for a one-way street. We agreed to study if a two-way street made any sense.

After business meetings in Bangalor, we went to New Delhi. The Taj Mahal is in a small agricultural town, Agra, that is three hours from Delhi. We were asked if we would like to take a day for GG to take us to the Taj. We went to the Taj. Needless to say, it was an awesome experience.

On the way to Agra, GG needed to stop and pay a road tax between states. We were told to keep our doors locked and windows rolled up because beggars would come up to the car. At one stop we saw an Indian boy walking up to the car with a burlap bag. He sat down by the car and dumped a cobra out of the bag. He needed to poke the cobra in the nose to wake up the snake so that it would perform while he played a flute. After the performance the boy wanted to be paid but we did not roll down the window.

A few months later Dale Weigel, Holland Agriculture Manager, and Kish Patel, Heinz Fairlawn Factory Manager, joined me for my second trip to India for an in depth agriculture review.

India: Jerry Shoup, Dale Weigel and Kish Patel

When we arrived at the cucumber fields, it appeared as though the whole village showed up to welcome the Americans to their village. They placed floral leis around our necks. The villagers also gave us a fruit drink. I was concerned about drinking it but was told it would be an insult if I did not drink it and so I drank it. We satisfactorily completed the agriculture review.

The final step was to complete sea container shipping tests to Grand Prairie, a Heinz Distribution Center in Texas. These tests were successful. GG was qualified.

I prepared a presentation to give to the Pickle Business Unit Team in Pittsburgh concerning GG co-packing some of our grocery spears to alleviate some of our seasonal labor shortages. Unfortunately Heinz Corporate stopped the project because they did not want country of origin labeling "made in India" on the Heinz label.

We decided to explore Brun Foods as a co-packer. Located in Colima, Mexico, the Brun family was well known in Mexico. They owned Pepsi-co franchises in Mexico. They had decided to diversify and go into the cucumber pickle business.

Brun had a unique business model whereby they grew cucumbers in green houses. This allowed them to control growing conditions and grow many crops of cucumbers throughout the year. Bill Snyder, Holland Engineering Manager, and I went to Colima to evaluate this opportunity. Heinz Holland had a good pickle market share in southern California. Our plan was to compare the delivered cost from Holland to Los Angeles to the delivered cost from Brun through the port of Manzanillo, up the Pacific coast to LA. The delivered cost from Brun was unacceptable.

⑰

HEINZ WATERFRONT WALKWAY 2008

One morning in 2003, I got a call from Soren Wolff, Holland City Manager. Soren asked me if Heinz would consider giving to the city and citizens of Holland access to the more than 1,800 feet of Heinz Holland Lake Macatawa shoreline to build a walkway. He explained that this walkway would extend the current Kollen Park shoreline walkway from the Boatwerks restaurant around the Heinz shoreline to 16th street.

Soren Wolff

I was keenly aware of Heinz's beginnings in Holland in 1897. I was overwhelmed with the thoughts of what this would mean to the community and the goodwill opportunity it represented for Heinz. I sat there thinking, "Soren you are asking me if Heinz would give back to the City of Holland and its citizens what they gave to Heinz 106 years ago to start a business in Holland, that being access to the Lake Macatawa shoreline." What a way to come full circle. I told Soren that I would do everything within my power to make this happen.

I called my boss, Dan Nolan, and he jumped onboard. We discussed the fact that Heinz does not use the shoreline. As a matter of fact having the shoreline maintained as part of an agreement with the city would be a huge benefit for Heinz. In the past, the Holland factory had been forced to spend capital to place rocks along the shoreline to prevent erosion from undermining the shoreline cliff and roadway around the factory to the warehouse shipping docks.

Shoup and Nolan Walkway plaque

The water levels of Lake Michigan and Lake Macatawa cycle up and down periodically. When cycling up, erosion can be a huge problem.

We were able to gain agreement from upper management at Heinz Corporate and with Heinz Legal to proceed with the walkway project. The Holland City Attorney worked with Heinz Legal to develop an "Easement in Perpetuity" to erect the walkway. I remember meeting with Mayor McGeehan and we were discussing a name for the walkway. The Mayor asked what I wanted to call it. I said "The Heinz Waterfront Walkway".

Former Mayor Al McGeehan

Over the next five years the Heinz Waterfront Walkway was designed and erected. According to state law, if a Corporation makes a gift of monetary value like the "Easement in Perpetuity" to a city then the city must give back to the corporation something of equal value. As a result, the City of Holland deeded to Heinz city property adjacent to the current Heinz property.

I worked with Mayor Al McGeehan and City Councilwoman Nancy De Boer to plan the dedication program and ribbon cutting ceremony on October 11, 2008. The dedication ceremony included U.S. Congressman Pete Hoekstra and State Senator Wayne Kuipers.

Mayor Nancy De Boer

Heinz Waterfront Walkway

Walkway ribbon cutting

HEINZ WATERFRONT WALKWAY 117

Shoup walkway presentation

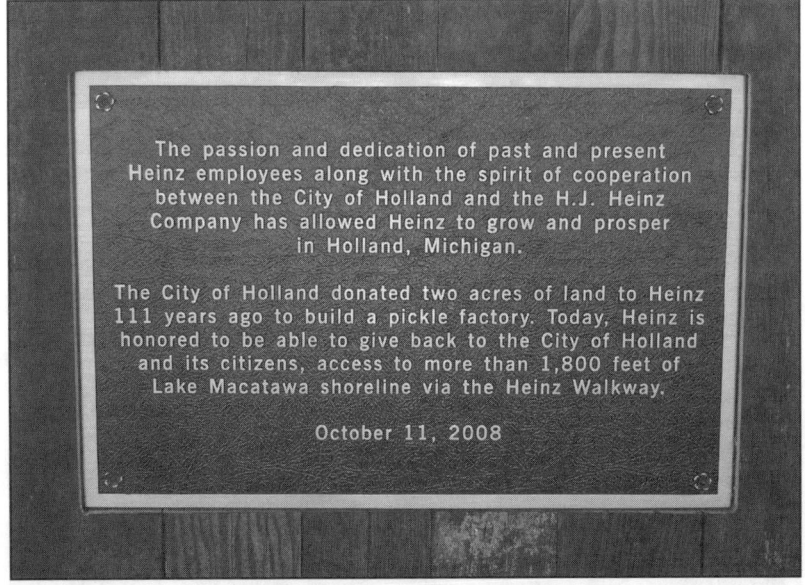

Walkway dedication plaque

AMERICAN MANUFACTURING

Walkway entrance from 16th St

18

HEART POWER

One of Henry J. Heinz's guiding principles was what he called "heart power." In the late 1860s, when he started his company, horses were important; however, one of his famous mottos was, "Heart power is stronger than horsepower." To Mr. Heinz, heart power was doing everything you can to make employees feel good about the workplace. To me, this meant capturing the hearts of all team members by treating them with dignity and respect and by acknowledging that what they do is an important part of the total team effort. I have indelible memories of heart power interactions with Heinz team members and members of the community.

I believed that it was important for me to communicate with all Heinz team members concerning factory performance in terms of productivity, safety and quality. I also wanted to give periodic business updates concerning brands which Heinz Holland served. I gave these updates during lunch periods.

I also gave annual State of the Factory presentations to all team members. Many of our team members spoke English as a second language. I knew that if I was going to communicate with all team members I needed to present in both English and Spanish. I am not fluent in Spanish although I took Spanish in college. I do know some words and how to pronounce them even though I may not fully understand their meanings. I prepared my presentations in English and worked with bilingual team members to translate them into Spanish, then practiced the Spanish versions with them.

I remember the first time team members heard me speak in Spanish. There were both English and Spanish speaking members in the lunchroom. I presented in English and when finished immediately spoke in Spanish. I will never forget the look on all team member's faces. I know I captured many hearts that day.

Tina was a wonderful elderly lady who was a widow who lived in a small house adjacent to the Heinz property. The house no longer exists. It was in the same location where the Heinz Waterfront Walkway entrance from 16th street is located.

Some years ago before the Heinz Waterfront Walkway was erected, we learned that Tina and her husband planted a marriage tree on their property. Unfortunately, one of our vinegar storage tanks leaked vinegar which ran down onto their property and killed their marriage tree. Once we learned about this, from that day forward, whenever Tina needed help with a plumbing problem or any other issue, Ryan Hunderman and our maintenance department took care of the problem.

Tina called me at home one day and told me how much she appreciated what Heinz did for her for many years. She told me she was 82 and just wanted to tell me that, since she was getting older. I assume she wanted to tell me because she did not know how much longer she would live. Not too long after that she passed away. Her attorney called and said that she named Heinz in her estate to have the first right of refusal to purchase her property and wanted to know if Heinz was interested in purchasing the property.

I said that we did not need the property. I thought to myself, I hope the city purchases the property. It would be best for Heinz if the city owned the land next to Heinz. Fortunately, the city did buy the property; the purchase facilitated the erection of the Heinz Waterfront Walkway.

I remember one of our department heads had an altercation with a forklift driver which resulted in some shoving. According to our work rules both parties must be terminated in a situation like this. This was a difficult situation. There had been previous shouting situations as well.

The normal process was to escort a terminated employee out the gate. I told him he could clean out his desk and walk out by himself.

A few years later, he called me and said you will never know how thankful I have been because you did not escort me out of the factory, embarrassing me in front of my department.

I remember walking around the production lines with a few supervisors. We walked up to one of the line leaders. The line leader looked at me and said "I bet you do not even know my name." I said "Your name is Debbie Gray and I know you are an important team member here." She had a surprised look on her face.

Luana Wise was one of the most dedicated and loyal team members that I have ever known at Heinz Holland.

She loved Heinz and everyone knew it. She was a wonderful person whom everyone liked and admired.

Unfortunately she developed cancer, which spread. I remember going to the hospital to see her. She was there with her family. Her son Joshua was there. Joshua was a U.S. Army veteran and she idolized him. He was everything to her.

I remember her saying to me that she could not believe that I came to see her. She thought that, since I was the manager, I must have a very busy schedule. She was overwhelmed that I took the time to come and see her. She passed away in January 2010. Joshua called me and said he would like to take me to see her grave marker. He said Heinz was on her marker. Luana's sister, Kristi Kiekintveld, told me that when Joshua and her went to get a grave marker for Luana they did not know what to put on the marker and so since her whole life revolved around Joshua and work at Heinz they decided to put Heinz on the marker.

Luana Wise Grave Marker

I cannot remember what was happening at the time but I could not go with him. I did search Pilgrim cemetery in Holland and found her grave marker. The Heinz keystone was on the bottom right corner and cucumbers on a vine in the upper left corner. To put this on her grave marker was such a wonderful memorial to her because they knew she loved Heinz and Heinz captured her heart.

Our team made a concerted effort to improve the appearance of and feelings for Heinz in the community. For example, the consolidation of all Michigan grower grading, receiving, and brining created dirt and dust and brought approximately 100 farmer trucks into Holland daily during the summer tourist season. This resulted in considerable traffic congestion and the perennial pickle and vinegar smells in Holland during the green cucumber season. The rumors of Heinz factory closing not only affected Heinz Holland employees but also feelings in the community about Heinz.

Everyone I ever met in Holland either worked at Heinz or had a relative or knew someone who worked at Heinz. This fact only seems reasonable given Heinz Holland is the only company in the history of Holland to span three centuries, from 1897 to 2013, under the same ownership and product line.

Given the rich history of Heinz in Holland we knew we needed to improve the image and feelings about Heinz with our employees and everyone else in the community. I did everything I could think of to be a good ambassador for Heinz in the community. The outsourcing of grower grading and receiving and the removal of all the old wood tanks and conversion to fiberglass tanks improved the appearance and reduced the perennial smells.

We made a concerted effort to improve the landscaping, appearance and cleanliness of the factory. Then, we worked with the Holland City and Heinz Corporate to make the Heinz Waterfront Walkway a reality; that was icing on the cake. We did make a difference and we all are proud of it. I think this also falls under the category of heart power.

HISTORY CHANNEL MODERN MARVEL PROGRAM ON ACIDS – VINEGAR

Heinz got a call from the History Channel concerning a Modern Marvels segment on acids. The acids to be included in the segment were sulfuric acid for cleaning metals, nitric acid for use in making explosives, and acetic acid, which is the main acid in vinegar. The Heinz headquarters called and asked if I would work with the production company developing the Modern Marvels segment.

I thought this would be an interesting project and started discussing the segment with the film production company. You may have seen this Modern Marvels segment on acids. I have received many calls and comments from those who have seen it.

To start the filming process, I wrote down every detail of the process steps used in making vinegar. The filming production company then reviewed the process steps and asked questions about the process.

We then agreed upon the date for the filming company to come to Heinz in Holland to film the segment on acetic acid. The filming company was at the Heinz Holland factory for about 10 hours to film what would be a 15 to 20 minute segment on acetic acid in vinegar.

This was a difficult filming session for me with many re-takes. The filming company editors were aware of the processing steps in making vinegar, which I had already outlined for them. Off camera they asked me specific questions about how vinegar was made which they thought the viewing public would be interested in hearing.

On camera, I discussed the answers in a way that didn't look like I was answering questions. I had to appear to be talking directly about the process steps. This was difficult for me.

My grandson told me that a high school chemistry teacher in Zeeland, MI, which is near Holland, included in his class a video of the Modern Marvels segment on making vinegar. There is a lesson to be learned in the conversion of ethyl alcohol to acetic acid when making vinegar. I was very pleased that the segment was being used in a high school chemistry class.

20

DELMONTE SPINOFF

Project Millennia was a global initiative to restructure Heinz for growth in the new millennium. Operation Excel took a huge step toward this global restructuring.

The impact of Excel upon the Holland factory was that new products and increased production of existing Holland products were transferred from the Pittsburgh factory to the Holland factory. As a result, Excel restructured the Pittsburgh factory to be a soup and baby food plant.

Excel set the stage for the Delmonte spinoff. Heinz spun off the Pittsburgh factory including its soup, baby food, and private label businesses to the Delmonte Corporation. Heinz also spun off its Starkist seafood factories and brands to Delmonte. All Pittsburgh Heinz headquarters' marketing, product development, and sales force personnel supporting these businesses became Delmonte employees.

In the past, the Heinz fiscal year ended on April 30th. In May, at the start of the new fiscal year, Heinz invited all management personnel to Pittsburgh for a yearend review and to discuss key goals and initiatives for the upcoming year. The management personnel invited to the meeting included those who worked in Pittsburgh and those from locations outside Pittsburgh, including the factories.

I remember being in a large auditorium when the VP of sales summarized the Delmonte spinoff. He made a statement which shocked me. He said, "Jerry Shoup, I know you are out there. Unfortunately the lawyers who developed the spinoff contract forgot to include private label pickles. Since Heinz no longer has support personnel here in Pittsburgh, we have recommended exiting the private label pickle business. If you want to retain this business, then you will have to manage it".

What a shock. Our private label pickle business for Wal-Mart, Aldi, Save-A-Lot, and Certified Groceries represented 30% to 40% of our remaining pickle business. To lose the business would seriously jeopardize our factory's viability.

As a result, Mike Mooney and Evan Dawdy from our quality assurance department took over product and package development functions. Ed Balint and Phil Wesorick took over the finance and bid pricing functions. I became the private label sales manager, calling on Wal-Mart, Aldi, and Save-A-lot. We decided to no longer bid on the Certified business since it was a small business. We took on these responsibilities in addition to our day jobs because we could neither afford nor justify hiring additional staff.

PRIVATE LABEL PICKLE SALES MANAGER

Bill Fatika, a former sales manager for Heinz who had called on Save-A-Lot in the past, agreed to train me. He was very helpful, especially for introductions to the various buyers for Save-A-Lot.

I remember my first sales call to Wal-Mart. I flew into Bentonville AR, Wal-Mart's Headquarters, on a small plane. The lady sitting next to me asked "What are you selling?" I did not respond right away. She then said, "Everyone on this plane is selling something." I said, I'm selling pickles. What are you selling?" She said, "Ladies Lingerie".

When we de-planed, we were herded into an area to undergo a SARs screening. This was at the time of the SARs scare. SARs is a respiratory disease which, if I am remembering correctly, was thought to come from China. A Wal-Mart nurse wanted to make sure we did not travel outside the U.S. in the last several weeks.

After the SARs screening, we were directed to the waiting area for a call to see the appropriate Wal-Mart buyer. When I finally did see the buyer, he wanted to know what I was going to do for him today. I was prepared to give a brief update on the cucumber crop but he was not interested. If I remember correctly, I believe we held the price for the various Wal-Mart pickles for the upcoming year, but that apparently was not acceptable. He wanted me to reconsider the pricing.

Sales calls with Aldi were much more professional. I learned quickly that an Aldi pickle salesperson should not ask the Aldi buyer if that person would like to go to lunch. Aldi buyers do not have any contact with sales people outside of the Aldi headquarters office in Batavia, IL. Sales calls with Save-A-Lot were more cordial. Going to lunch was acceptable. It was expected that your service history was good and your pricing was no higher than any other bid.

A few years after I became the private label sales manager, we were fortunate to get the Jack Daniel's barbeque sauce product transferred into Holland. Production of Jack Daniel's barbecue sauce required us to exit the private label pickle business, due to changes we needed to make to the production line to accommodate the Jack Daniel's product. These changes eliminated our ability to produce grocery spears for private label customers. We did install a small grocery spear hand pack line for Heinz grocery spear production.

22

MILLENNIA IMPACT UPON HOLLAND

The purpose of Millennia was to restructure the global corporation for future growth. This certainly did trickle down to Heinz Holland, transforming the pickle and vinegar factory into a condiment and sauce factory, greatly increasing the profitability of our product portfolio and the viability of the factory.

I strongly believe that our management and RWDSU team enabled Heinz Holland to survive and that we laid the foundation for future growth of Kraft/Heinz as evidenced by the introduction of the iconic Kraft brand Grey Poupon into Holland.

23

COMMENTS FROM OUTSIDE AND INSIDE HEINZ

I sent the book proof out to several non-Heinz members of the Holland community and others and former Heinz Holland employees with whom I worked during my Heinz career. I wanted to get their comments about the book and the changes made during my time managing the Holland Factory and Agriculture operations.

Comments from Outside Heinz

Former Mayor Al McGeehan commented "Jerry, your new book will surely showcase the history of Holland's oldest production facility, it's people, it's product(s) and it's commitment to excellence. But, in addition it will tell the story of how well meaning folks from the private and public sector can partner to achieve some amazing results. It was this partnership that first brought Heinz to Holland over a century ago and then a decade ago brought about the now ever popular Heinz Waterfront Walkway. It was my pleasure to collaborate with you and your colleagues in Pittsburgh to bring about the walkway. Let the history on that be made perfectly clear. The walkway would have been a great idea and remained only an idea, had it not been for the voice, guidance and support from Jerry Shoup. You made it happen."

Current Mayor Nancy De Boer commented "Amazing, Jerry! Holland is forever grateful to you for your wisdom and sacrifice as such a noble leader of the Heinz plant and the community. Thanks so much for preserving the principles for future generations!" Nancy.

Holland author and historian Randy Vande Water commented "Hello Pickle King! A fascinating story and you told it well. The anecdotes of your 38-year career are priceless and illustrate well the countless decisions you made. Congratulations on your book. It is well done throughout, content and pictures."

Holland Museum Registrar Rick Jenkins commented "Thank you so much Jerry, for writing the book and sharing it with us."Holland businesswoman Darlene DeWitt commented "Thank you! I'm anxious to read the book cover to cover! I've lived on the south side of Holland almost all my time here in Holland. Driving past Heinz and smelling the vinegar from the pickle production during the summer months was just a normal part of my daily life. Thank you for all you did for the company, the product, the workers, and the positive impact Heinz had on and in our Community!"

Former Holland City Official who held several high level positions in Holland City government Jodi Syens commented " I have been reading your book and just finished it tonight. First of all, I want to say congratulations on such an impressive career. I realize there is much I didn't know about your background, your training, and your extensive experience, and I am grateful for it. We don't always take the time to recognize the amount of effort that goes into making the food we eat safe and delicious. What an extraordinary effort on the part of you and your team over the years! I also want to thank you for highlighting the positive relationship between Heinz and the City of Holland in so many ways. Obviously, I'm somewhat biased but I think it is such a positive thing to point out the great things that can happen when business and local government work together. Obviously, the results in Holland support this. Finally, I want to applaud you for your tireless efforts to continue to make Heinz a real presence in Holland during challenging financial times. It really shows how much you cared for the company and your employees, and I thank you for being a manager with a true heart for people. Heinz and Holland, Michigan are two things that should never be separated, and you have played a significant role in making that possible.

Again, congratulations on the publication of this book, and thank you for the privilege of allowing me to read it."

President of Swanson Pickle Company Ravenna, MI John Swanson commented "Our family company has been a pickle supplier to the Heinz Holland Plant for over fifty years, many of those with Jerry at the helm. Jerry was a pleasure to work with. His knowledge of the industry, inner workings of the plant and its workers along with the understanding of the heartbeat of Heinz is why the plant became such a viable part of Heinz USA and a bedrock of the Holland community. Jerry is a man of integrity who has always had passion for life and his work with Heinz. Jerry started with Heinz and ended his career with Heinz, something that is almost unheard of anymore. I miss those days.

Comments from Inside Heinz

Former General Manager of Manufacturing in Pittsburgh Headquarters Dan Nolan commented "Jerry and the Holland team created a unique culture at the Holland Factory. Their stewardship as owners of the Factory and support of the Holland community allowed them to react to an ever changing business climate. It was by this ownership culture and attitude that the factory weathered multiple restructurings, possible closures, and business downturns. Under Jerry's leadership, the Holland team and the Holland community transformed a seasonal pickle factory into a year round Heinz condiments operation."

Former Heinz Fremont, OH Factory Manager Jerry Kozicki commented that the title of the book is wrong. It should be titled "Henry, Pickles and Me. Jerry also said that he wants to play himself in the movie.

Former Heinz Fairlawn, NJ Factory Manager Kish Patel commented "as you contemplated retirement, you and I walked back from the Pittsburgh Pirates baseball game across the Bridge and we both agreed you still had the Holland Factory in your DNA and you were not quite ready to retire yet. Also when you do retire there will be a great legacy you will leave for both the Holland Factory and the Holland Town."

Former Heinz Holland Manager of Production Larry Heckel commented "I had the privilege of supervising or managing several Heinz factories and had built a reputation as the guy who could turn around struggling production operations. So I wasn't surprised when asked to transfer to the Holland Factory, which was definitely not on upper management's "good" list. I accepted the challenge and figured improving Holland would be the next successful step in my career. But after arriving my big surprise was meeting Jerry Shoup for the first time. My initial thought was, "who the heck is this guy?"

I listened to the explanation of his being appointed acting factory manager and agreed it made sense, at least in the short term, due to the complexity of the pickle business. Upper management may have thought they took a gamble by placing Jerry in charge. But results under his leadership replaced concern and the day soon came when it was announced that Jerry was no longer "acting." He was the manager. It was hard work and long hours, year after year. Plans were developed and implemented which transformed Holland Factory from a problem riddled pickle facility to a modern state of the art factory capable of producing a variety of products. I am proud of the team that saved Holland Factory and very proud of the guy who gave it direction and all his effort, my boss and friend, Jerry Shoup.

Former Heinz Holland Factory Controller Ed Balint commented "Over the years the Heinz Holland team, under Jerry Shoup's leadership, had many different Headquarters bosses. They all had different management styles but regardless of style most were very supportive of the Holland Team's dedication, passion for excellence and credibility resulting from consistent delivery of positive results. One of the most memorable and highly respected Headquarters bosses was John Turyan. John's management style was somewhat unique and in some respects ahead of the times. John's philosophy was exemplified in a course which he strongly believed in and supported titled "Value of the Person". A major tenant of the course was to treat everyone with "Dignity and Respect". John expected/

required that all team members embrace and follow this simple principle. The Holland Team did embrace it and was able to accomplish amazing results that many did not believe possible. John said on numerous occasions that Holland Factory had the potential and all the synergies to become the "Model Factory" in Heinz North America and even Heinz globally.

Former Assistant Quality Assurance Manager Evan Dawdy commented "I read your book and really enjoyed the journey back in time. I appreciated your listing of the various team members and their contributions. It was clear that you attributed your success to the efforts of a talented and dedicated team. As I always said, you were the visionary that never ignored an opportunity. I phrased it differently... I think I said "You never saw a business that you didn't like". The olive tanking endeavor was one I didn't know about but it was clearly consistent with your ability to see business where no one else did. The book demonstrated to me that a large part of the plants success was the consistent team of talented managers, workers, and union leaders. The team trusted you to be fair, honest, considerate, and appreciative. You recognized the contribution of your team members, their sacrifices, and their accomplishments. Success and trust made the work enjoyable and the team feel like family. Thank you for that! Evan"

AFTERWORD

I believe that when Heinz started to lose U.S. pickle market share in the 1960s and 1970s, Heinz decided it was not worth spending a lot of money to defend that market share. The Pittsburgh headquarters started to focus its research and technical resources on more profitable businesses, such as ketchup. I certainly agree that this was the right thing to do given the growth and success of Heinz.

The pickle business was still important for Heinz, given the tradition of the company and the fact that many consumers still thought of Heinz as a pickle company. I believe Heinz needed to find a champion who could stay up to date on all aspects of the pickle business. This would allow the Pittsburgh headquarters' research and development, marketing, and other personnel to focus their resources on Heinz's more profitable and growth businesses. I happened to be in the right place and time to play this role, which I thoroughly enjoyed for 38 years.

REFERENCES

H. P. Fleming, "Mechanism For Bloater Formation in Brined Cucumbers" Paper no. 6236 of the journal series of the North Carolina Agricultural Research Service, Raleigh, N.C. 1980

Randall P. Vande Water, "Heinz in Holland: A Century of History" Color House Graphics 2001

Robert C. Alberts, "The Good Provider H. J. Heinz and his 57 Varieties" Houghton Mifflin Company Boston 1973

Robert P. Swierenga, "Holland Michigan from Dutch Colony to Dynamic City – Volume 1 Holland Shoe Factory" pp 800 – 803 Van Raalte Press